Total quality in construction projects

achieving profitability with customer satisfaction

Ron Baden Hellard

Quality management must be applied to, and achieved on, the building project. Auditing of people, as well as paper systems, plays an important role in achieving customer satisfaction — the purpose of total quality management.

T| Thomas Telford, London

Published by Thomas Telford Services Ltd, Thomas Telford House, 1 Heron Quay, London E14 4JD

First published 1993

Distributors for Thomas Telford books are
USA: American Society of Civil Engineers, Publications Sales Department, 345 East 47th Street, New York, NY 10017—2398
Japan: Maruzen Co. Ltd, Book Department, 3—10 Nihonbashi 2-chome, Chuo-ku, Tokyo 103
Australia: DA Books and Journals, 648 Whitehorse Road, Mitcham 3132, Victoria

A catalogue record for this book is available from the British Library

Classification
Availability: Unrestricted
Content: Guidance based on best current practice
Status: Refereed
User: Clients, consulting engineers, architects, project managers, contractors, specialist subcontractors, building clients

ISBN: 0 7277 1951 3

Typeset in Great Britain by MHL Typesetting Ltd, Coventry

Printed in Great Britain by Galliard (Printers) Ltd, Great Yarmouth

When you can measure what you are speaking of, and express it in numbers, you know that on which you are discoursing. But if you cannot measure it, and express it in numbers, your knowledge is of a very meagre and unsatisfactory kind.

Lord Kelvin

FOREWORD

SINCE THE Second World War, in line with advancing technology, we have seen increasingly specialised fragmentation
in the construction industry, contributing to divisive and
adversarial procurement, contractual and operational
procedures. We also see the need for improved co-ordination
to serve better perceptive clients who are critical of the quality
of both the service and product provided by the industry.

This last point has to some extent been met by the recognition
of management as a key function at every level: first, by the
importation of (not always appropriate) ideas and techniques
from manufacturing, and then by the development of the best
of this knowledge and experience to meet the special needs of
the construction industry.

During this time there have also been new concerns about
quality, which has been defined as 'exact fitness for intended
purpose'. Ideas about quality control (mostly to do with
specification and inspection) and quality assurance (certification
to meet national BS 5750 and international ISO 9000 quality
standards) have again been imported from manufacturing, and
not surprisingly both have been found wanting in a project-
based industry where design is typically separated from
construction.

Thus there is a need for new unifying perceptions of
management and quality to reduce separatism and conflict, and
to improve professional service with a strong client orientation.

The titles of his earlier books, *The management of architectural
practice* (1964) and *Managing construction conflict* (1988), show that
Ron Baden Hellard is uniquely qualified to present in this book
a well structured, clearly argued and readable case that total
quality management (TQM) is such 'an idea whose time has
come' to assist the UK construction industry reorganise to meet
the challenges of competitive international markets on the
threshold of the 21st century.

He leaves the reader, whether expert or amateur, in little
doubt that total quality management has a substantial part to
play in the better management of the industry overall and in

all its parts that must work as a team to produce a quality project.

I have found the book stimulating and enjoyable to read — I commend it to all the industry's professionals.

Ian Dixon, CBE
Chairman, Construction Industry Council

ACKNOWLEDGEMENTS

THIS BOOK would not have been possible without the experience gained by Polycon over nearly 40 years' involvement in consultancy in the construction industry, initially in the design of buildings and project management of building projects and then as management consultants engaged on design and productivity studies with and for clients, building product manufacturers, contractors, subcontractors and professional design firms. This experience has been honed more recently by work on specific quality management assignments with a similar range of clients. In addition, several packages have been developed by the author's Polycon colleagues.

The author has drawn substantially on these packages to present a logical and reasoned argument for the use of total quality management in all the constituent parts of the industry, and on construction projects, and then for the development of the technique of auditing to maintain quality in them.

Sincere thanks are therefore due to countless individuals whose words and actions have provided the evidence on which this work is based. Specific acknowledgement is due to Johnnie Beale, David Leggett and the late Martin G. Paterson, who did much on the development of the auditing package. Diana Porter and Angela Newland are thanked for word-processing the manuscript, a procedure which has made the writing of this book much easier and more enjoyable.

The author wishes to thank the British Standards Institution for permission to reproduce Fig. 14(a) and Appendix 2; the Department of Trade and Industry for permission to publish extracts from the report of the Henderson Committee; the Chartered Institute of Building for permission to reproduce Fig. 12; the European Federation for Quality Management for award criteria related to the European Quality Award; Messrs Tozer Gallagher for permission to reproduce Fig. 19; and Elsevier Science Publishing Co. Inc. for permission to reproduce Fig. 7, which is taken from 'A preface to motivation theory' by A. H. Maslow, *Psychosomatic Medicine*, vol. 5 (copyright the American Psychosomatic Society Inc., 1943).

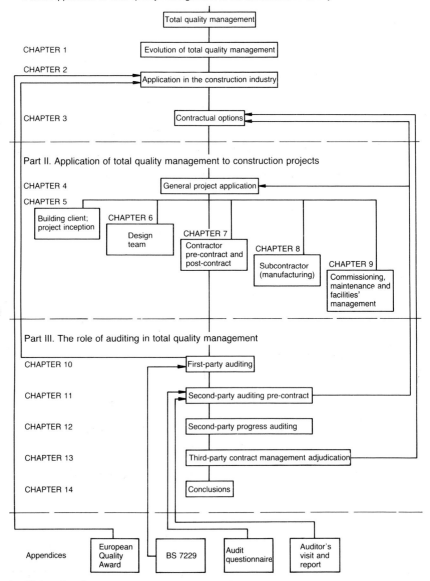

Auditing in the construction process: flow diagram showing chapter interrelationships

CONTENTS

LIST OF FIGURES AND TABLES

INTRODUCTION

A UDITING HAS long been accepted as a practical and worthwhile tool to monitor and report on the financial control, status and position of a company or organisation practising the concepts of total quality management.

However, finance is only one of a whole range of management areas that make up a modern business organisation. The same is true of any large organisation concerned with public or commercial endeavours, or with the provision of services such as health or the many social and technical services provided by local authorities and public sector industries.

Management is a whole process. Finance is but one spoke in a wheel that must have a hub and a rim if it is to perform effectively in any modern sphere of activity.

Quality management auditing extends the process of auditing to cover all the spokes in the wheel of a corporate operation. Its development in the second half of the 20th century through quality assurance of the production process into total quality management has been such that it is now seen as a third industrial revolution. It has been forecast that by the end of the 20th century there will be only two kinds of commercial operation — one which has embraced and practices total quality management and the other which will have gone out of business! This view accepts the premise that the quality company is the profitable company and that the pursuit of quality adds value and reduces the cost of the product. This increases customer satisfaction and so develops long-term success for the organisation.

The development of quality control as a company-wide industrial technique to improve a company's performance through its products began in the USA in the 1920s and was much developed in Japan in the 1950s through American management consultants Deming, Juran and Crosby — now considered to be the gurus of quality — working with Japanese industry. Today, the success of Japanese products across the world has forced the USA, Europe and the UK to re-examine their position in an increasingly competitive and quality-conscious world. Progressive manufacturing industry, whether large or small, has come to recognise that quality management is a must for survival.

In the UK the first quality assurance standard, *Quality systems* (BS 5750), was for product development, production, installation and servicing, and was published in 1979 when the Minister responsible for UK industry was Michael Heseltine. He realised the significance of this standard for manufacturing industry and urged its adoption. In 1992 he returned to this same office with the additional title of President of the Board of Trade, and is now concerned to put total quality management into high profile for British industry.

In the interim, the quality assurance concept has had great success in developing quality systems within manufacturing firms. In ten years over 12 000 firms have been included on a register of those who have had their systems 'third-party certified' by an accredited certification agency.

The success of BS 5750 across the board has misled many people in the construction industry to follow too closely the procedures set out in this series of standards and to be unaware of the differences needed for effective practice in the construction process. This results from the lack of understanding of the principles and, above all, of the overriding philosophy without which the practices of quality management will not have any long-term significance or success in achieving their real purposes within an organisation — the improvement of its overall performance.

This comes about partly because of the pressure resulting from third-party certification. So much has this been the case that many people in the construction industry believe that quality assurance is synonymous with paperwork systems — it is not — and that paperwork systems are synonymous with BS 5750 — they are not — and that BS 5750 is synonymous with certification — which it certainly is not!

Total quality management, however, has a tremendous part to play in maintaining and improving the performance in both the design and construction parts of the industry. To be successful, total quality management must start at the top, and in the construction industry, which is essentially project-related, this must be with the building client.

Auditing has a major role to play in ensuring the effective and continuous improvement of construction projects. One purpose of this book is to enable the reader to fully understand the role of auditing in the construction process. Perhaps more importantly, it aims to clear away the confusion that has arisen from the pressure to introduce third-party certification. This procedure, excellent in adding value for the production lines of manufacturing industry and also very useful in commercial situations where repetitive services are provided, is inappropriate and not cost-effective for most of the construction process.

Before considering the role of auditing it is therefore important to understand how the philosophy of total quality management affects the construction industry and the one-off construction project.

DEFINITIONS

Definitions of quality and related terms

Assessment register (DTI QA register). Register maintained by the Department of Trade and Industry of firms which have been audited and assessed by an independent certification body as being of assessed capability.

Certification. There are three types of third-party certification.

(a) *Certification of a quality management system.* This applies to a firm involved in providing a product or service. The firm is then authorised to use the NACCB symbol in its literature.

(b) *Product-conformity certification.* In this case, in addition to certification under (a), the product will have been assessed and tested to ensure that it conforms to the specification by the British Standards Institution (BSI) kitemark, the BSI safetymark, and Certification Authority for Reinforcing Steel (CARES) schemes.

(c) *Product approval.* In this case, in addition to certification as in (a) and (b), the product is assessed for performance in use (e.g. British Board of Agrément Certificate).

Certifying body. An independent certification body, which will itself have been assessed as a firm capable of reliably working to specification and having a management control system that will comply with BS 5750. In the UK a recognised certifying firm will be accredited by the National Accreditation Council for Certification Bodies (NACCB).

Defect. The non-fulfilment of intended usage requirements.

Non-conformity. The non-fulfilment of specific requirements. Non-conformity may apply to one or more quality characteristics of a service or elements of a quality system.

Product or service. This may be

(*a*) the result of activities or processes (tangible product; intangible product, such as a service, a computer program, a design, directions for use)

(*b*) an activity or process (such as the provision of a service or the execution of a production process).

Quality. The totality of features and characteristics of a product or service that bear on its ability to satisfy stated or implied needs.

Quality assurance. All activities concerned with the attainment of quality — a process designed to increase confidence in a product's or service's ability to achieve the stated objectives.

Quality audit. Examination to determine whether quality activities comply with planned arrangements and whether these are implemented effectively to achieve objectives.

Quality control. The operational techniques and activities which together sustain the product, service or quality to specific requirements.

Quality loop; quality spiral. Conceptual model of interacting activities that influence the quality of a product or service in the various stages ranging from the identification of needs to the assessment of whether these needs have been satisfied.

Quality management. That aspect of the overall management function that determines and implements the quality policy.

Quality management system. The total management system of organisational structure and responsibilities, activities, resources and events which together provide procedures and methods of implementation that ensure the organisation is capable of meeting the quality requirements.

Quality plan. A document setting out the specific quality practices, resources and sequence of activities relevant to a particular contract or project.

Quality policy. The overall quality intentions and direction of an organisation, as formally expressed by top management.

Quality system review. A formal evaluation by top management of the status and adequacy of the quality system in relation to quality policy and new objectives resulting from changing circumstance.

Registered firm. A firm whose products or services have been certified after assessment by an accredited third-party organisation as being capable of consistently satisfying a need.

Reliability. The ability of an item to perform the required function under stated conditions for a stated period of time.

Service. *See* Product or service.

Traceability. The ability to trace the application or location of an item or activity by means of recorded identification. Traceability may be applied to a product or service, the calibration of measuring equipment or calculations and data relevant to a product or service.

See also BS 4778 *Glossary of terms used in quality assurance (including reliability and maintainability terms)* (from which some of these terms have been taken).

Definitions used in relation to construction-related quality

Quality. For construction, the needs must be defined by the client. The inclusion of services is pertinent to construction, where both designers and contractors supply services as well as the product (i.e. the completed work). The quality of these services is vital, not only in meeting the client's requirements, but also in completing to time and budget (function, aesthetic, cost and time — FACT).

State-of-the-art audit. An initial audit done before a quality management programme is instigated. It is concerned as much with the 'people culture' as it is with the existing paperwork and operating procedures.

Building process. The overall series of linked activities beginning with the identification of need and conception of the project and ending with the occupation and initial use of the construction after its commissioning.

Project.   The word used to indicate the overall activity being undertaken as opposed to the use of the term 'contract'.

Contract.   The legal document and framework under which the various parties carry out their work on the project.

Parties included in a building project

Client representative.   The person appointed by the client either from within his own organisation or specially appointed to be the single point of communication between the client and all parties to the contract, particularly for contractual instructions. He would therefore normally be named for this purpose in all contracts subsequently entered into by the client.

Project manager.   A project manager may be appointed in addition to a client representative. He will need to have his role defined and his authority, responsibilities and accountabilities indicated in the terms of his appointment and then subsequently in all contracts entered into between the client and other parties to the contract. The use of a project manager does not eliminate the need for a nominated client representative but it will probably reduce the scale and scope of the activity of the client representative.

Design professional.   An individual or firm providing a professional design service connected with the project in accordance with the code of conduct of one or other of the professional institutions. This covers architects, structural engineers, civil engineers, building surveyors, landscape architects, building services engineers and quantity surveyors. Any of these may be employed on a design contract, which may be individual for the technical area of design or may be a collective appointment to a firm offering a multidisciplinary service. The firm distinguishes between the role as a professional and a similar role in which members of the disciplines are directly employed by a general or specialist contractor.

General contractor. A firm undertaking the principal activity of construction. It implies overall control of the construction works, the contract for which normally empowers the contractor to enter into subcontracts with 'domestic' subcontractors on his own terms, or with specialist sub-contractors on terms which may be defined by the client's agents, whether or not the specialist subcontractor is 'nominated' by the client.

Specialist contractor. A firm providing a specialist part of the building such as heating services, lifts or curtain wall envelopes. As is the case for the general contractor, the specialist contractor may include technical design as well as manufacture and installation on site, sometimes including the necessary commissioning activities.

Craft contractor. A firm supplying particular craft skills such as decorative plasterwork, plastering, asphalting or classical stonework. Craft contractors are generally small firms and employed as domestic subcontractors.

1 The evolution of total quality management

This chapter examines the main contributions (and contributors) to the evolution of the concepts of scientific management and quality over the past 100 years. It recognises the work of the first work-study pioneers, focusing on time and method techniques for improving productivity, and on the work related to human motivation by Maslow, McGregor and Herzberg. It discusses how quality considerations have progressed from inspection to total management, and the contributions made by Crosby, Deming, Juran and others. All of this has led to an awareness that total quality management has the important cultural philosophy that customer satisfaction is inseparable from business goals. It is shown how this company-wide ethos requires management systems developing around a massive training programme in all areas and at all levels, from managing director to junior operative, to develop the culture that recognises that everyone in the organisation has a customer either within the company or outside it.

TOTAL QUALITY management has been described as the third industrial revolution, which has emerged from a rapid development in the third quarter of the 20th century. It has had two main streams contributing to its development: that of scientific management and that of quality.

The scientific management stream

The main stream of evolution of scientific modern management resulted from the study of work at the end of the 19th century by Taylor and Gilbreth, who applied their mind to the methods, machines and materials involved in time study. They recognised that the development of techniques for improved performance required first the development of proper measuring techniques: techniques to measure input against output, and the balance of requirements against resources, so that productivity measures could be established and through them better ways found to improve performance within an organisation.

Time was seen as an important measurement device, and finer measurements of performance in both production and people systems were necessary. New terms and concepts needed to be identified so that the results could be communicated and so that an understanding of them could be developed by other people and, perhaps for the first time, across disciplines.

Work study produced terms such as 'Therblig' (Gilbreth backwards), a time unit for micro measurement; 'standard performance', from 'standard rating' relaxation allowances; predetermined motion time systems; and 'activity sampling', leading much later to 'critical path networking', with particular application to one-off projects. All this reflected that management systems are technical systems for the direction of men and materials towards a stated objective.

In the construction industry, the application of materials embraces also the operation of machines and methods, all of which have financial implications, which have long been audited.

The direction of men to achieve required objectives may be assisted by financial incentives, but the early work in time and motion study led to conflict between unions, representing the direct interests of labour, and management, who were seen by the unions as the champions of the shareholders who received the greater benefits derived from a more profitable operation.

Studies of management on a more social plane were the subsequent work of Drucker, Maslow, Herzberg, McGregor, Urwick and others, in which human motivation was added as an important element in the satisfactory conduct of commercial and business organisations.

The quality stream

The addition of the quality stream into this evolution of management began perhaps 20 years later in manufacturing organisations. The first move was the inspection of finished goods before they were despatched from the manufacturing plant to prevent, wherever possible, the delivery of imperfect or unsatisfactory products. This inspection process received particular attention in armaments factories during the First World War, when people unskilled in factory work were involved in the production of shells and explosives. The explosion of shells in the breach of artillery weapons did little harm to the enemy but a great deal of harm to the gunners!

Later, the inspection process applied at the end of the production line was recognised as being too late, as correction still involved reprocessing, with unnecessary cost. *Quality control*, the application of control procedures during and at the start of the production process, was then developed. Later still, *quality assurance*, the systematic examination of documentation through a paperwork management system, further improved the cost-effectiveness and reliability of the manufacturing process. The work of Deming, Crosby, Juran and, in particular, Feigenbaum in the development of mathematical theories of statistical process control, and many others in the renaissance of Japanese industry after the Second World War, contributed greatly to the quality management philosophy. The contribution of these quality-stream gurus has emphasised the importance of monitoring, through audits, to the never-ending process of improvement.

An overview of the development of scientific management and quality considerations is shown in Fig. 1.

The Deming contribution

Dr Deming's advice is now in great demand. He conducts four-day seminars in quality management from which two exercises — the red bead exercise and the funnel experiment — have become classic illustrations in quality training.

The red bead exercise is a simulation of a factory. Willing workers are directed to make white beads. The process involves dipping a paddle into a mixture of white and red beads. The paddle has 50 depressions and extracts that number of beads from the mixture. No matter how hard the workers try, they never succeed in producing white beads without a red one mixed in. In the course of the exercise those involved learn several lessons, including the following.

- Willing workers are doing the best they can. Exhortations and threats cannot improve quality.

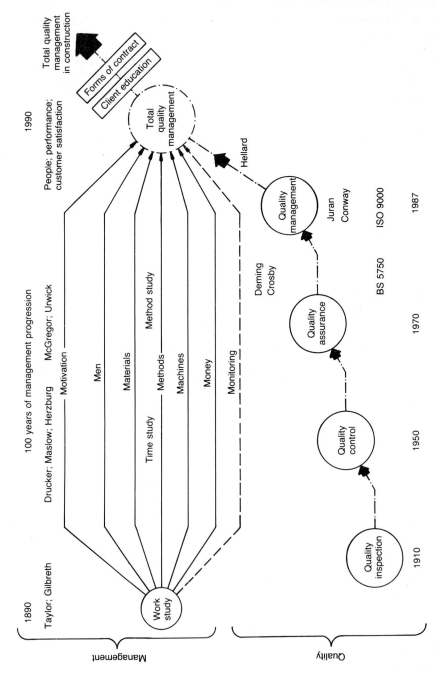

Fig. 1. The evolution of technique, practice and philosophy leading to total quality management in construction in the 1990s

- Improvements will come only by changing the process. This is management's job.
- Variation is a part of every process. It must be understood to be controlled.

In the funnel experiment a marble is dropped through a funnel over a target. If it comes to rest away from the target, the location of the funnel is changed according to a set of rules, and another marble is dropped.

One set of rules moves the funnel away from the target the same distance as the marble, but in the opposite direction. This illustrates the attempt to overcome variation by adjusting a process against the direction of error. For example, if a machine produces a rod longer than target, it would be adjusted to make shorter rods. The result of this tinkering is shown by the funnel experiment to double the variation in the product from that of a process left alone. The lesson is again to understand variation and reduce it by process changes rather than increase it by tinkering.

Another set of rules moves the funnel over the location of the marble after each trial. This compounds the errors and ultimately drives the variance to infinity. The lesson illustrates Deming's contention that as worker trains worker more and more errors are introduced into the process. It is therefore management's responsibility to provide training and retraining in the proper methods of doing the job.

Deming also claims that quality benefits the worker, as shown in the Deming chain reaction

improve quality — costs decrease — productivity improves — better quality and lower prices capture the market — business survives and grows — more jobs are created.

Deming lists 14 points for corporate improvement (Table 1). He also has what he calls his seven deadly diseases (Table 2); to these he adds quick-fix gadgets, and gurus with example but no theory.

Admirable as these theories are, their application to construction will require much re-assessment of the theory and certainly of its application before the philosophy can be successfully used. (This will be no problem to the reader of this book: it will be a few chapters yet before the role of auditing and its practical tools are set out!)

Table 1. Deming's 14 points for corporate improvement

1. Create constancy of purpose for improvement of product/service. [Long-range needs, not short-term profits.]
2. Adopt the new philosophy; outlaw mistakes and negativism; develop teamwork.
3. Cease dependence on mass inspection; quality comes not from inspection but from improvement of the process; build-in quality. [The only way it can happen in construction projects!]
4. Stop awarding business on price-tag alone; purchasing departments must seek the best quality, not the lowest price, and try to achieve it with a single supplier in a long-term relationship. [Of great significance in construction — although particularly difficult.]
5. Improve constantly and forever the system of production and service; management's job is to look continually for ways to reduce waste and improve quality.
6. Institute training and retraining — learning is as important as producing. [Where does this put management-only organisations in construction?]
7. Institute leadership — in place of management by dictat — helping, coaching everyone to do a better job.
8. Drive out fear — fear of asking questions or taking responsibility.
9. Break down barriers between staff areas — different departments often work not for the same goal but in competition. [The norm for construction projects through the contractual nexus?]
10. Eliminate slogans, exhortations and targets (e.g. 'zero defects'). [Zero defects will never be the case with one-off projects!]
11. Eliminate numerical quotas — they usually guarantee inefficiency. [Even more meaningless in construction.]
12. Remove barriers to pride of workmanship (e.g. merit rating.)
13. Institute a vigorous programme of education and retraining — self-improvement on a broad basis.
14. Take action to accomplish the change; a top management team with a plan of action is needed to carry out the quality mission by leadership — support is not enough. [Whose role in the construction project?]

Table 2. Deming's 'seven deadly diseases'

1. Lack of constancy of purpose — management by departmental conflict (the construction industry norm)
2. Emphasis on short-term profits (the norm in construction)
3. Evaluation by performance, merit rating, or annual review of performance — they destroy teamwork and nurture rivalry
4. Mobility of management — it undermines the long-term changes needed for quality and productivity
5. Running a company on visible figures alone — the key figures (like the effect of a happy customer) are unknown
6. Excessive medical costs
7. Excessive costs of warranty (construction!)

Other contributors to total quality management

Philip Crosby believes that the problems of poor quality and performance have been brought about by poor management, and particularly by the 'quality professionals'. He urges that the 'soft' or attitudinal part of quality must be addressed and that this must begin with management commitment and leadership and with the concept that all work is a process. Each process can be improved when it is managed by the following four absolutes of quality management.

- Quality is defined as conformance to requirements, not 'goodness'.
- The system for causing quality is prevention, not appraisal.
- The performance standard must be zero defects, not 'that's close enough'.
- The measurement of quality is the price of non-conformance, not indices.

Applying these four absolutes will require both management systems and individual and technical tools for the company and its managers. Some are very relevant to the construction process but others need substantial re-orientation.

J.M. Juran is the author of many practical handbooks on managing quality. His philosophy is summarised in the 'Juran trilogy': quality planning, quality control, and quality improvement.

His approach is frequently empirical, and was learnt from as much as taught to the Japanese; he suggested that the reduction of the number of defects on a job produced 'happiness', and that the improvement of quality drove down cost and increased market share. Hands-on leadership, massive training investment — as illustrated by the JIT (just in time) development — and the insistence on annual improvement in quality were all seen as part of the planning, control and improvement trilogy.

The Juran approach has been the driving force for the almost fanatical attention to quality in the USA. The successful companies applying it have all had outstanding leadership who were well informed, who set dramatic but realistic quality goals and who provided the necessary resources for the goals to be met. They have also had a strong focus on the customer, and they have looked at themselves and their competition, identified the 'best in class' in both practices and processes, and have established benchmarks to compare and continuously implement and improve what they have learnt. The total quality management systems introduced are nevertheless unique to each company rather than related, except at the lowest and simplest level, to external standards.

Feigenbaum, who emphasised mathematical theories of standard process control, defined quality costs as the sum of prevention costs, appraisal costs, internal failure costs and external failure costs. In construction arbitration, litigation and claims handling represent failure costs.

Genichi Taguchi is noted for his emphasis on the reduction of variation and the creation of robust designs (i.e. designs which continue to perform well as the use environment varies). His contributions include improved methods for statistical design of experiments to determine causes of variation. He formulated 'loss functions' to quantify the adverse economic effects of variation.

Taguchi's contributions are often explained by considering a design hierarchy: system design, parameter design and tolerance design. System design creates the means to accomplish some mission, and American designers are strong in this area. Parameter design is concerned with the specification of the system components: this is a Japanese strength, and a Taguchi speciality. Tolerance design, the setting of limits on specified values, is done equally well by both Americans and Japanese.

Robert Townsend, not generally thought of as a guru in this field, wrote a book, *Up the organisation*, published in 1970, which recognised many of the points made later for total quality management. He preached rebellion against mindless rules which accumulate in all organisations. He suggested that managers call their own offices to see what impressions a customer gets when he calls. He noted the importance of leadership, the need for a manager to be a coach, and the general under-utilisation of people in an organisation. He asks, 'If you can't do it excellently, don't do it at all. Because if it's not excellent it won't be profitable or fun, and if you're not in business for fun or profit, what the hell are you doing here?'

Norman Augustine wrote *Augustine's laws*, a book describing the American aerospace industry. It is amusing until you realise that he is not exaggerating. 'It costs a lot to build bad products' (cost of quality), 'most of our problems are self-imposed' and 'rules are no substitute for sound judgement' are telling quotes from his work.

These two separate management streams began to come together in the 1980s, and by 1990 a whole concept of total quality management had emerged and a philosophy and principles had begun to be established.

Total quality management is now seen as a means of survival for the company, a means of increasing profit for the shareholders, a means of ensuring long-term job security for the workers, and a means of providing the workers with greater job satisfaction through the development of a greater sense of involvement and team-working within a shallower management structure.

Kaoru Ishikawa, a Japanese contributor to the philosophy of total quality management, developed the theme as a thought revolution for management, essentially a moral concept of honesty and integrity in terms of providing quality to customers. 'When Quality Control is implemented falsehood disappears from the company', he says. In his book *What is quality control? The Japanese way* he emphasised

- leadership by top management
- education from top to bottom
- action based on knowledge and data
- teamwork; elimination of sectionalism
- customer focus
- prevention of defects by eliminating root causes
- elimination of inspection
- use of statistical methods
- long-term commitment.

Ishikawa contends that 95% of a company's problems can be solved by using seven basic tools. He contends that any engineer can look at six of them and understand immediately how to use them; the seventh, a control chart, is more complex. His tools are

- stratification (i.e. use of flow charts)
- Ishikawa diagrams (fishbone cause-and-effect charts)

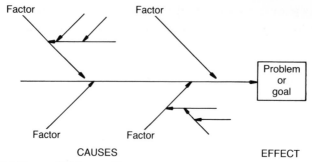

Fig. 2. Ishikawa fishbone diagram

- check-lists
- histograms
- Parato charts
- scattergrams
- control charts.

The flow chart shows all the inputs and outputs of all operations in a process; it should also show 'feedback', which should be a part of every process. This tool provides an understanding of the process being studied. Charting the real process often shows it to be very different from what its managers thought it to be.

The Ishikawa diagram (Fig. 2) shows the factors which cause an effect in a problem or on progress towards a goal. The bones of the chart can be any of the factors, but the four Ms — Method, Manpower, Materials, Machinery — are a good starting framework. Each bone can have any number of subordinate bones. The purpose of the chart is to isolate the factors which can then be worked on to solve the problem.

Towards a common philosophy

From all of these contributions, a start can be made on developing a common philosophy. In essence, total quality management is a philosophy where everyone wins: the company, all shareholders, the worker and the customer.

Not all the quality gurus agree on all the points made by others; indeed, there are some direct conflicts in application. The following are some examples.

- *Forms of recognition.* Should workers who serve on improvement teams be rewarded with money? Is it better to acknowledge their contributions by public thanks or a certificate? Or should the satisfaction of making an improvement be enough?

- *Use of hoopla.* Some initiatives in total quality management have been kicked off with much razzmatazz and parades. Banners and coffee mugs with total quality management slogans abound. Crosby would recommend these: Deming might consider them empty exhortations. Who is right, or does it depend on the organisation?
- *Setting goals.* Is this an essential activity or a counter-productive exercise? Motorola's so-far-successful 'six-sigma' programme has a yearly target for defect reduction. Deming might point out that an easy approach to victory would be to change the definition of defect (something not unknown in many bureau-cracies). Does it take a target to sustain energy, or is it an invitation to play numbers games?

However, there is general agreement that total quality management is an approach to managing work based on

- the analytical evaluation of work processes
- the development of a 'quality' culture
- the 'empowerment' of employees

all for the purpose of continuous improvement of the product or service.

So, agreement on the principles and issues of total quality management can be summarised as being concerned with

- customer satisfaction
- management leadership to create a quality culture
- improvement of processes, not 'motivation' of people
- education and training (job skills and total quality management tools at least, but for management a broad 'education' is also needed)
- defect prevention in lieu of inspection
- use of data and statistical tools
- developing a team approach — horizontally (between departments) and vertically (chief executive to lowest-paid employee)
- continuous improvement.

Total quality management is not

- new
- a programme (as opposed to a process or philosophy)
- a quick fix or magic solution
- spiritual guidance
- a slogan campaign
- a Japanese invention
- a suggestions programme
- a substitute for discipline and dedicated effort.

It also is definitely *not* easy.

In implementing total quality management, the chief executive officer must lead the way. He cannot delegate leadership, nor can he ever put quality on the back burner, or total quality management will never mature under his agency.

If a union represents the employees, they must become partners in the implementation of total quality management. They should be involved at the start and all information relating to total quality management must be shared with them. The union should define its own role in total quality management, but it should be separate from established grievance and bargaining procedures.

A massive training programme is necessary to deploy the vision and create a quality culture. Job skills and basic quality concepts must be provided to all employees. Those involved in improvement teams should have available to them training in

Table 3. Culture changes needed

From	To
Bottom-line emphasis	Quality first
Just meet specification	Continuous improvement
Get product out	Satisfy customer
Focus on product	Focus on process
Short-term objectives	Long-term view
Delegated quality responsibility	Management-led improvement
Inspection orientation	Prevention orientation
People are cost burdens	People are assets
Sequential engineering	Teamwork
Minimum-cost suppliers*	Quality-partner suppliers*
Compartmentalised activities	Co-operative team efforts
Management by edict	Full employee participation

* A very big change for construction subcontractors.

Table 4.   The Boeing model for process improvement

1.   Assign process owner and define boundaries
2.   Flow-chart the process
3.   Establish effectiveness and efficiency measures
4.   Determine process stability
5.   Implement process improvements
6.   Validate improvement
7.   Document improvement

group dynamics and in total quality management analysis tools of interest.

Improved quality must never be seen as a threat. When an improved process needs fewer man-hours to perform, the people released should be considered an asset available for other uses and guaranteed continued employment.

Total quality management has been described as a culture change. Culture changes are needed in many areas, and Table 3 could be used as a statement of goals and as a check-list to gauge progress. It is taken from a programme of the US Air Force, where total quality management is seen as equally valid as in commerce or industry.

Much of the work on which total quality management is based has focused on statistical analysis and process control, pioneered in the 1920s by Walter Shewhart and subsequently developed by Deming and later by Homer Sarenson. For the construction industry these mathematical approaches are of less value than elsewhere, for the processes are infrequently repeated, if at all. However, the Boeing seven-step method can help improve construction processes (Table 4).

Assessing and auditing quality efforts through quality awards
USA
So strong has the total quality management movement become in the USA that a national award, the Malcolm Baldridge National Quality Award, has been instituted, with the prize awarded by the President. There is general recognition that competition for this prize has helped industry forward by requiring self-assessment (first-party auditing) against the award criteria.

The Malcolm Baldridge award was established in 1987 to stimulate American companies to improve quality, to recognise achievements, to establish guidelines for self-evaluation, and to make available information from successful organisations. The award is not without detractors, who argue that the Baldridge criteria may not be the only way to world-class quality. They note, for example, that the emphasis on participatory management assumes that other management approaches would not work. Nevertheless, the Baldridge criteria provide a *de facto* standard for judging the quality efforts of an organisation and are the basis for most other criteria, including the more recent European Foundation for Quality Management award (see appendix 1). The Baldridge criteria and the relative weights given to each factor are shown in Table 5. The factors are further divided into subfactors as shown in Table 6.

It can be readily seen that if the organisation is a firm of consulting engineers or architects, some sections in Table 5 may need a much stronger emphasis and many subsections in Table 6 will need substantial additions and deletions. Clearly, until appropriate benchmarks can be established for organisations in the construction industry — not to mention projects — the auditing and assessment will be difficult. Nevertheless, the target is there and construction cannot, must not, be left out.

Each of the subfactors in Table 6 is further divided into 2−4 areas, and at this level also changes are needed. As a result, Baldridge applications take significant effort. They also take

Table 5.   Baldridge criteria and relative weights

	Factor	Weight: points
1.	Leadership	100
2.	Information and analysis	70
3.	Strategic quality planning	60
4.	Human resource utilisation	150
5.	Quality assurance of products and services	140
6.	Quality results	180
7:	Customer satisfaction	300
		1000

time. But the whole process gives the preparer a deep insight into his organisation's quality efforts.

Already, non-manufacturing organisations are involved in the USA, for from this original manufacturing framework has followed the Quality Improvement Prototype (QIP) award. This award was established for two reasons: to recognise government organisations that have successfully adopted total quality management principles and thereby improved the

Table 6. Subdivision of Baldridge criteria, with weighting

Factor and subfactors	Weight: points	Factor and subfactors	Weight: points
Leadership		*Quality assurance (continued)*	
Senior executive leadership	40	Process quality control	20
Quality values	15	Continuous improvement	
Management for quality	25	of processes	20
Public responsibility	20	Quality assessment	15
		Documentation	10
Information and analysis		Business process and	
		support service quality	20
Scope and benchmarks	30	Supplier quality	20
Analysis of quality data	20		
		Quality results	
Strategic quality planning			
		Product and services	
Strategic quality planning		quality results	90
process	35	Business process,	
Quality goal and plans	25	operational, and	
		support service quality	
Human resource utilisation		results	50
		Supplier quality results	40
Human resource			
management	20	*Customer satisfaction*	
Employee involvement	40		
Quality education and		Determining customer	
training	40	requirements and	
Employee recognition and		expectations	30
performance		Customer relationship	
measurement	25	management	50
Employee well-being and		Customer service standards	20
morale	25	Commitment to customers	15
		Complaint resolution for	
Quality assurance		quality improvement	25
		Determining customer	
Design and introduction of		satisfaction	20
quality products and		Customer satisfaction	
services	35	results	70
		Customer satisfaction	
		comparison	70

Table 7.   Quality improvement prototype criteria and weights

Factor	Weight: points
Quality environment	20
Quality measurement	15
Quality improvement planning	15
Employee involvement	15
Employee training and recognition	15
Quality assurance	30
Customer focus	40
Results of quality improvement efforts	50
	200

efficiency, quality and timeliness of their services or products; and to use the quality improvement prototypes as models for the rest of government, showing other agencies how a commitment to quality leads to better services and products. The QIP evaluation criteria and weightings are shown in Table 7.

UK

The Henderson committee was set up by the Department of Trade and Industry in 1992 (under the chairmanship of Sir Denys Henderson, Chairman of ICI) to study the feasibility of improving the performance of British industry through an award for total quality management. The committee adopted the following as a working definition

'Total quality management is a way of managing an organisation to ensure the satisfaction at every stage of the needs and expectations of both internal and external customers — that is, shareholders, consumers of its goods and services, employees and the community in which it operates — by means of every job, every process being carried out right, first time and every time.'

The committee's task was to study the guiding philosophy behind the Malcolm Baldridge National Quality Award (established in the USA in 1987), and the European Quality Award set up by the European Foundation for Quality Management in 1991. It reported that there is now universal agreement that getting quality right is of the greatest importance to all businesses and other organisations providing services.

UK needs and applications

British industry faces greater commercial competition than ever before from its counterparts in Europe and other developing countries, as well as newly industrialising and developing nations. This competition sometimes arises from the drive to maintain leadership in the market for sophisticated and technically advanced services; in other cases it comes from the challenge of economies with lower labour costs. At the same time the expectations of customers, employees and investors have risen and technology is advancing with increasing rapidity.

Against this background, failing to satisfy customers' needs and expectations, or failing to do so right first time, has been acknowledged as costing manufacturing companies somewhere between 10% and 30% of their sales revenue. In some service companies the percentage can be even higher.

Companies acknowledge that quality management provides a framework for action to fully satisfy customer requirements by setting out an approach to

- identify customer needs and expectations
- set standards consistent with customer requirements
- control processes and improve their capability
- establish quality standards
- make clear management's responsibility for setting quality policy, providing leadership and equipping people to achieve quality
- enable and empower people at all levels in the organisation to act for quality improvement.

In the UK these total quality management principles are being adopted also by hospitals, health and social services, public utility undertakings and others.

Many have studied the work of Deming and Juran and also adopted salient points from Taguchi, Townsend and Augustine, building in many cases on the earlier work of McGregor and the Tavistock Institute.

People-participation and standards

In the UK a significant new development emanating from the Training and Enterprise Directorate of the Department of Employment has been the establishment of an 'investors in people' standard providing a third and significant strand to the evolution of total quality management. This picks up and develops the earlier work of Maslow, whose view was that the reaction of human beings depends fundamentally on the extent of their ascent up the basic pyramid representing the hierarchy of human needs. The approach recognises that the behaviour of an individual, in all interests or work situations, is the individual's total response to all motivating forces. The total quality management implication of these theories is examined in more depth in relation to their application in the construction industry in chapter 2.

The national standard for effective investment in people recognises four principles. Top management

- must show public commitment
- must make effective investment in its people and plan a 100% training culture
- must develop an action plan to train and develop its people
- must establish benchmarks against which appraisals are made to enable an evaluation of its investment in its people to be assessed.

Team-working

Total quality management has emerged as an overall approach to modern management where team-work is seen as the dominant feature for improving corporate performance.

Recognition of team-working will need attention to

- employee recruitment
- selection and promotion
- rewards and recognition
- skills assessment and training
- employee involvement and empowerment
- tranformation of managers into coaches.

The seven principles that will drive the team-work approach are as follows

- the approach — management-led
- the scope — company-wide
- the scale — everyone is responsible
- the philosophy — prevention not detection
- the standard — right first time
- the control — cost of quality
- the theme — ongoing improvement.

It will be seen that these new concepts linking principles and the philosophies from which they have emerged require measurements of a more sophisticated nature and these will require benchmarks to be established against which auditing can take place.

Conclusion

By 1992 various defitions of total quality management had been developed by various authorities. The British Quality Association defined it as

'a corporate business management philosophy which recognises that customer needs and business goals are inseparable. It is applicable within both industry and commerce.

'It ensures maximum effectiveness and efficiency within a business and secures commercial leadership by putting in place processes and systems which will promote excellence, prevent errors and ensure that every aspect of the business is aligned to customer needs and the advancement of business goals without duplication or waste of effort.

'The commitment to TQM originates as the chief executive level in a business and is promoted in all human activities. The accomplishment of quality is thus achieved by personal involvement and accountability, devoted to a continuous improvement process, with measurable levels of performance by all concerned.'

A draft addendum to *Quality vocabulary*, ISO 8402, defines it as

'a way of managing an organisation which aims at the continuous participation and cooperation of all its members in the improvement of the following

— the quality of its products and services
— the quality of its activities
— the quality of its goals

to achieve customers satisfaction, long term profitability of the organisation and the benefit of its members, in accordance with the requirements of society.

'It involves every department, function and process in a business and the active commitment of all employees to meeting customer needs. In this regard the 'customers' of each employee are separately and individually identified.'

Some uses of these new auditing practices with many of the spokes of the total quality management wheel will have direct application in the construction industry; others will require substantial modification for the project situation, with its nexus of contracts essentially different from the simple buyer and seller relationship that provides customer satisfaction for a manufactured product. The involvement of the external and ultimate customer, the building owner, must be brought in to the team concept. The team concept itself will also have many differences of application within the construction industry. Chapter 2 considers these differences of application for the construction industry and project.

2 Application of quality as a management process in the construction industry

This chapter looks at the evolution of the process of building from the simple two-party relationship of builder and customer to the present-day arrangements for highly sophisticated projects with their need for complex management structures and procedures. It relates the project's management needs to quality management principles, and in particular to the problems of conflict inherent in the design and construction of one-off projects. It therefore also examines the implications of human motivation and behaviour in conflict situations, calling on the earlier work of Maslow and McGregor; and how third-party involvement of adjudicators rather than quality assessors can help in the final stages of delivering project quality through the continuation of effective team-working involving the client, the design team and the building contractor.

THE PRESENT-DAY client has a choice of routes through the building process (Fig. 3). How this system has evolved is examined below.

Construction is project-related for one-off projects. Initially the relationship between the builder and the client (or group requiring the building) was simple, and similar to the buyer-seller relationship conceived by the International Standards Organisation's standards for quality assurance. These customary practices and the relationships between the different groups have now become complicated by the organisational or contractual groupings of the different functional elements. Multidisciplinary professional firms sometimes combine with general building contractors to carry out specific projects as joint ventures, and it is often now the norm on any large project for specialist suppliers and subcontractors to combine with, or to be off-shoots from, larger main contractors in conglomerate financial groups (Fig. 4).

The ultimate in this situation was probably reached on the Channel Tunnel project, where ten general contracting firms in the UK and in France combined with several banks to become two entrepreneurial companies, indivisibly linked by the division of each share into two, with half held in each other's company, to construct the tunnel and its infrastructure and working parts. The parent company then placed contracts with the various contracting companies, who were initially

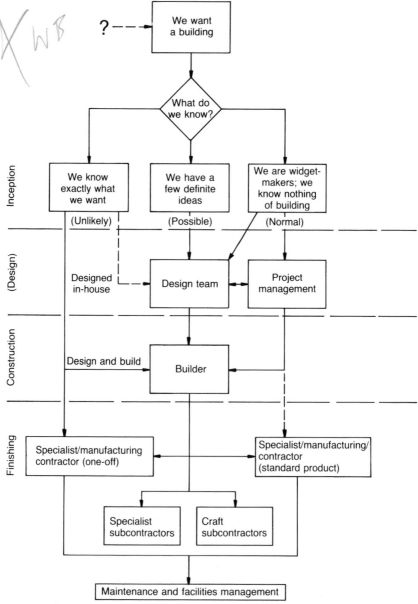

Fig. 3. The client's alternative routes through the building process

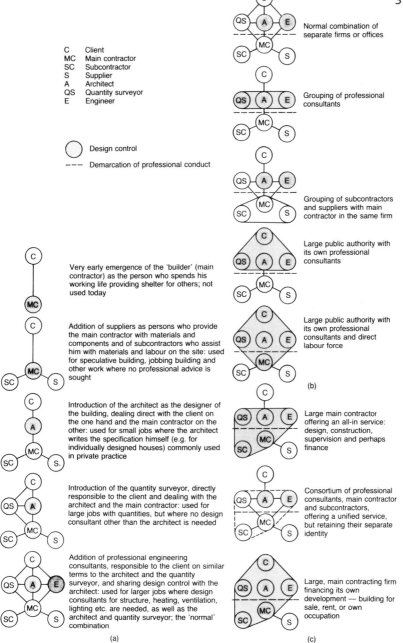

C   Client
MC   Main contractor
SC   Subcontractor
S   Supplier
A   Architect
QS   Quantity surveyor
E   Engineer

◯   Design control

---   Demarcation of professional conduct

Normal combination of separate firms or offices

Grouping of professional consultants

Grouping of subcontractors and suppliers with main contractor in the same firm

Large public authority with its own professional consultants

Large public authority with its own professional consultants and direct labour force

(b)

Large main contractor offering an all-in service: design, construction, supervision and perhaps finance

Consortium of professional consultants, main contractor and subcontractors, offering a unified service, but retaining their separate identity

Large, main contracting firm financing its own development — building for sale, rent, or own occupation

(c)

Very early emergence of the 'builder' (main contractor) as the person who spends his working life providing shelter for others; not used today

Addition of suppliers as persons who provide the main contractor with materials and components and of subcontractors who assist him with materials and labour on the site: used for speculative building, jobbing building and other work where no professional advice is sought

Introduction of the architect as the designer of the building, dealing direct with the client on the one hand and the main contractor on the other: used for small jobs where the architect writes the specification himself (e.g. for individually designed houses) commonly used in private practice

Introduction of the quantity surveyor, directly responsible to the client and dealing with the architect and the main contractor: used for large jobs with quantities, but where no design consultant other than the architect is needed

Addition of professional engineering consultants, responsible to the client on similar terms to the architect and the quantity surveyor, and sharing design control with the architect: used for larger jobs where design consultants for structure, heating, ventilation, lighting etc. are needed, as well as the architect and quantity surveyor; the 'normal' combination

(a)

Fig. 4. The evolution of functional relationships and operational groupings of firms engaged in the building industry (based on work by Sir Roger T. Walters, *RIBA Journal*, February 1960): (a) historical development — combination of separate firms to form building teams; (b) the normal building team, and variations whereby functions are grouped together in one firm; (c) combinations assembled into larger units under one management control for larger projects

major shareholders for the construction of the project; but these general contracting companies had already entered into a variety of arrangements with dozens of firms, some of them their own subsidiaries and others professional consulting firms, for the design and supply of components or services connected with the project. Subsequently the parent company established a main company that will operate the tunnel when it is completed, and this company acted as the Client against the constructing contractors.

Quality assurance techniques were introduced but claims have proliferated at all levels. Total quality management on a project basis did not exist. Certainly the client and contractors did not see themselves as a single team pursuing a common objective throughout the project.

These complexities are still only part of the pattern of relationships that are created on any large project. Others are formed through the constraints or influences of people and organisations outside the client and the team assembled to design and construct his requirements. Figure 5 identifies those organisations external to the project team with whom the building owner, his architect or project manager must first come to terms to ensure that the client's requirements can be met.

The first phase in the overall process of building must always be establishing the requirements for the project — the essential standards against which performance can later be assessed by audit, either objectively against benchmarks or subjectively by 'customer perception'.

Getting the brief

Every construction project has four frequently-conflicting elements which must be established in 'the brief'. These four can be classified by the code word FACT

- Function — all the technical and physical requirements: space, servicing, internal relationship between the parts, access, egress and the like
- Aesthetic — the satisfaction of all the human and subjective aspects that will be enshrined in the end result — the modern equivalent of commodity firmness and delight

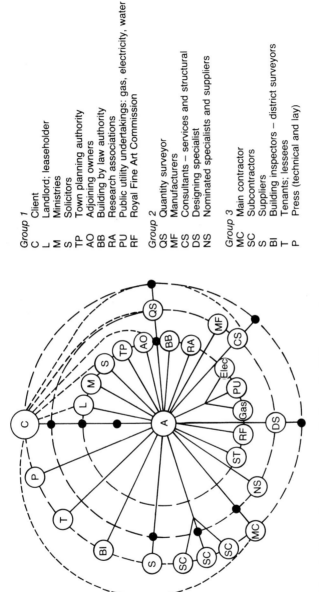

Group 1
C   Client
L   Landlord; leaseholder
M   Ministries
S   Solicitors
TP   Town planning authority
AO   Adjoining owners
BB   Building by law authority
RA   Research associations
PU   Public utility undertakings: gas, electricity, water
RF   Royal Fine Art Commission

Group 2
QS   Quantity surveyor
MF   Manufacturers
CS   Consultants – services and structural
DS   Designing specialist
NS   Nominated specialists and suppliers

Group 3
MC   Main contractor
SC   Subcontractors
S   Suppliers
BI   Building inspectors – district surveyors
T   Tenants; lessees
P   Press (technical and lay)

Fig. 5. Organisation network for a building project

- Cost — both capital and running costs; perhaps better expressed as lifetime cost of the project
- Time — the logistic requirements for commercial completion and occupation; in some cases (e.g. a short-term exhibition project), even this can be the most critical requirement in the client's brief.

Sometimes the requirements under each of these headings can, or should be, provided by the client, but sometimes they must be established from the person or authority that can exercise a modifying or even controlling influence over the matter. The influence of such external authorities is rarely complete or a matter of black and white.

Even when the control, such as with town planning authorities, has statutory power, the client's representative, project manager or design team leader (to use the terminology established by the British Property Federation's System of Organisation for Building Projects) can still bring about a satisfactory solution by negotiation, and the design may be the better for the challenge presented by these conflicting requirements.

However, it can clearly be seen that when so many people are involved in providing the criteria for the brief and so many technologies are involved in satisfying the requirements of the brief in a design solution, even before work begins on site, the whole situation is one where conflicts of requirements and differences in the resources available to satisfy the requirements abound. Total quality management must address these issues.

If all those involved are given the credit of wishing to perform their own tasks to the best of their ability, it can easily be seen that there is the need for a great deal of negotiation, which in its turn will demand excellent communication ability on the part of the person or persons responsible for finalising the design brief. Many of these negotiations will continue over weeks or months and will be interactive as between the client's stated requirements and the constraints imposed by the external parties, many of which will themselves be in technical conflict with each other.

At some point the design must be frozen to enable the second stage of the operations — the detailing of materials, with their consequent quality and cost implications — so that a tender can be obtained for implementing the design by construction.

Implementing the project requirements

The degree of finality in the price obtained from the various contractors for the work required of them will, or should, depend on the extent of the firmness of the requirements stated in the tender. The greater the firmness, the less the contractor's risk; therefore the more competitive should be the price.

Whatever contract is obtained in terms of time and price, however, the external constraints, if not the client's own situation, are subject to the possibility of continuing change. Changes in the people or controlling power within the external authority can result in changes of policy, and these in turn can produce a different situation (which may be to the building owner's disadvantage or may produce a relaxation from which the building owner can benefit) and so produce changes to the brief or design requirements, which in turn can change design and so reflect on the contractual terms.

All this is perhaps to state the obvious, but it shows why building contracts provide for variations within the contract. In principle, every change does create a situation where the cost and time criteria are completely open to renegotiation. This requires considerable modification of the overall philosophy of total quality management and the procedures of quality assurance expounded in ISO 9000. The various standard forms of contract have grown in size so that they now prescribe what are essentially customs and practices for dealing with this situation. Because they are developed (and indeed argued over) within the industry, these customs and practices reflect more the conditions which are generally perceived to be prevailing within the industry than they do the legal environment outside the industry — or ISO 9000.

Construction has become ever more complex and technical, and the relationships and contractual groupings of those who are involved are also more complex and contractually varied.

Cultural change

Every culture, whether society, club, trade union, or professional, working or social group, develops customs and rules, written and unwritten, which govern its actions, often more than the regulations under which it was set up or continues to exist. Within the construction industry, architects, quantity surveyors, engineers, contractors and various specialists all have, in addition to their special technical skills, their own trade or professional customs and practices, some of which come together in an overall framework for building projects, but some of which do not.

Changes happen slowly, particularly in the older cultures. Change is generally mooted only when the old pattern can clearly be seen, within as well as outside the established group, to be essential to the continued well-being of that group. The changes proposed are generally such as to be acceptable to the members of the group, particularly when the pressure is from outside. Pressure for change and improved quality in the construction industry has come from outside by informed clients in industry and government, but hitherto has been for the assessment and auditing techniques of quality assurance and ISO 9000 developed in the manufacturing environment. This has been helpful in arousing interest. In manufacturing, the application of these techniques has led to the development of total quality management for improving the broader overall performance within the company but outside the production process: in construction, the broadening process must be further developed to affect the management of the project process.

Management definitions

As shown in chapter 1, total quality management has emerged from the scientific principles of Taylor and Gilbreth, supplemented 50 years later by the thinking of Deming, Juran and others that there is always a better way. Today management must combine mathematical and physical sciences with the human sciences of people and their relationships.

Management is about balance: the balance of requirements against resources, of income against expenditure, and of the needs of the group in the achievement of collective or corporate goals against those of the individuals involved in this overall achievement. It is also about maintaining the balance between the law on the one hand, and custom and practice on the other, while moving dynamically, through the direction of resources, human and others, to achieve a given objective.

Management is therefore concerned with motivation as well as with men, machines, materials, methods and money, and it is about procedures and practices. But total quality management is also about principles and philosophies, and in construction it is about developing these principles and philosophies to provide effective and profitable outcome to projects which must cross cultures.

There are as many definitions of management as there are pundits. The definitions perhaps reflect the definer's management style or attitude as much as his analysis of the subject. Definitions range from 'getting work done through people — a social process' to the far more authoritarian and mechanistic 'planning and regulating the operations of an enterprise in relation to its procedures and to the duties and tasks of its personnel'. The International Labour Organisation in 1937 defined management as 'the complex of coordinated activities by means of which an undertaking, public or private, is conducted', and 'organisation: the complex of activities the object of which is to achieve optimum coordination of the functions of the undertaking'.

The first definition, 'getting work done through people', was given to me in the mid-1960s by Dr J.F. Dempsey, who was then the Chairman of the Irish Management Institute. In 1936 he had joined Aer Lingus as the manager of the first Irish aeroplane company, with six people and a single de Havilland aircraft flying three times a week the 200 miles between Dublin and Cardiff. When the author met him he was still Chief Executive of Aer Lingus and had been instrumental in developing the company so that it became an international airline by the mid-1970s.

Those who have observed over the years the promotion of Aer Lingus will know that it has described itself as 'the friendly airline', and those who have flown Aer Lingus will know that, whatever other attributes it has, the airline staff, now many thousands strong and flying many millions of miles every year on daily schedules to every continent and the most developed countries of the world, certainly reflect that adage. Jack Dempsey's attitudes, personality and influence were clearly reflected in the management of Aer Lingus through its formative and growth periods.

Another definition of management which seems to be appropriate to the construction industry is 'the direction of men and materials to a given end'. This carries implications of technique or scientific method, which springs from the origins of modern management through 'work study' as developed by Taylor, Gilbreth and others. These pioneers realised that the development of techniques to improve performance required first the development of appropriate measuring techniques.

Thus management systems are technical systems for the direction not only of men but also of materials towards the stated objective. In the construction industry the application of materials embraces also the operations of machines and methods, as these too have financial implications.

As for the direction of men, today a literate and highly numerate community have greater freedom of choice than their predecessors as a result of economic progress. The direction of men may be assisted by financial incentives, but motivation means more than just money for workers in the latter part of the 20th century.

Management in the construction process: project management

In construction there is an essential difference in the application of management principles, in that a construction project has a finite life. It has a beginning — the establishment of the need, in detail, in the client's brief; it has a middle — the design and development of the solution to that need; and it has an end — the implementation by the contractor of the solution by the physical construction of the building. After this, the project's people structure and the other physical resources which have been employed are broken down and redeployed on other projects under other managers for other organisations. But the fundamental principles still apply.

Total quality management must therefore produce a cohesive body of knowledge which is logically arguable, if not yet always scientifically provable, to achieve a balance and harmony through which the stated objective — generally the construction of the client's building — can be economically achieved.

Management functions, project requirements and quality standards

Whereas definitions of management are wide with regard to both its technique and its philosophical limits, there will be more general agreement on the functions and principles which must be practised to achieve success. These functions can be analysed for the manager as shown in Table 8.

But how do all these apply to the construction industry? How do they apply in a dispute? Who is the manager? Who are the managed? What are the manager's resources? And what is the project objective? — will this not vary depending on whether the point of view taken is that of the client (or building owner), the architect or the engineering consultant, or the contractor? Furthermore, what, if any, of this analysis applies to the contract — that is, the set of legal promises which the parties have entered into to bring about the fulfilment of the client's building needs?

The analysis should help to establish standards which are needed in setting up the management structure and performing the tasks required for the completion of the project by the 'Agreement', which has itself been evidenced by all the contract documents — that is, the drawings, specifications, schedules, networks and programmes, in addition to the printed standard JCT (Joint Contract Tribunal) or other form, or a variation on one of these. Those managing a building project need also to recognise a number of other fundamental principles of management and see that they are implemented in turn by those involved in the project. The requirements of ISO 9000 standards (but not the principles and philosophy behind them) will do little or nothing to clarify these project-specific requirements except where they relate to the manufacturer of products such as bricks, tiles and sanitary fitments used in the building.

However, there are other more important factors that need to be addressed, and audited against suitable benchmarks.

Table 8.   Functions of the manager

Planning
  Forecasting the future so that the enterprise may be in a position to
    continue to operate effectively
  Setting the objectives for the workforce under his control
  Defining the policies whereby these objectives will be achieved
  Establishing the outline procedures through which these policies will be
    implemented
  Preparing the budgets for the financial implications of those policies
  Scheduling the tasks involved in carrying out these policies

Organising
  Developing a system within which individual efforts can be harnessed
    together in the most effective way to produce overall momentum in the
    organisation
  Delegating his responsibility for particular areas of the operation and giving
    with it the authority to control these
  Establishing relationships between those involved at different levels in the
    organisation structure on a formal basis

Leading and motivating
  Making decisions which affect himself and other people
  Selecting those with whom he will carry out his tasks
  Training people who will have to work within the organisation
  Understanding people, their own mainspring, their personal objectives, self-
    motivation and will to work
  Developing the environment which will stimulate individual action and
    responsibility
  Establishing effective communication within and outside his organisation to
    ensure that confusion does not arise

Controlling
  Setting standards of performance
  Measuring standards of performance
  Assessing the results of progress
  Taking corrective action where this is necessary as a result of deviation
    from the standards

The client's role

In a building project, unlike many other management
structures, the top management is frequently, one might say
generally, amateur in that the task is performed only once, or
at best only infrequently, by the client, the prospective
building owner. The remainder of the management team, as
shown in Fig. 4, are all professional, and expert. However, it
is the client who must make key decisions on what, where,
when and with whom, and to what standards, the project is to
be built.

Then there is the further division between 'professional' consultants and 'commercial' contracting activity, with these groups performing functions in separate corporate organisations but coming together in a variety of arrangements to fulfil the client's requirements for the project.

It is the nexus of human relationships, which becomes a nexus of contractual relationships, that is at the root of many dispute situations, with which construction is rife. Far more disputes arise through failures in organisation and in communication between the different groups than result from failures of technology or materials, or even from the situations which develop from unforeseen events.

Before any functions can be carried out each manager at every level in the management pyramid must have defined for him his authority, responsibility and accountability — a fundamental requirement for a total quality management plan.

Initially, within the client organisation this concept must be used to establish and define the client's objectives for his building. These should be crystallised in the architect's brief. The brief should recognise the relative needs or wishes of the client in terms of function, aesthetics, cost and time.

The establishment of the brief requires skill and in this the client has usually not had the opportunity to develop the skills needed. The establishment of the brief must therefore be an interactive process between the architect or engineer and the client. The brief should not be too determined, as this could frustrate the architect's ingenuity and his ability to create an original, imaginative and effective solution in four dimensions — time being added to the spatial concepts.

The design ability of the architect or engineer can only be established and evidenced by reputation if either is to play a project management role. A pre-appointment assessment will be necessary to ensure that the architect or engineer has both an appropriate quality management system and the available resources. Part III examines these matters in detail (see also appendix 3).

Functional management responsibility: development of the project

Figure 6 illustrates as a functional management responsibility

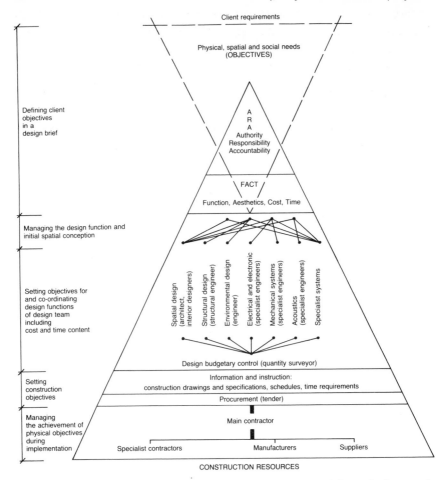

Fig. 6. Management responsibility pyramid for building design and construction functions: balancing requirements against resources — conflict or balance?

pyramid the development of the project from the initial concept into a detailed design for eventual implementation on site by the construction team.

Establishing the brief: whose responsibility?
The successful fulfilment of the client's needs requires progress through several distinct phases, at several of which pre-appointment audits will be desirable.

First, getting the brief is an interactive process between the client and the designer or design team. The client's physical, spatial and social requirements must be analysed into what might be considered in management terms as the design objectives, all of which should be expressed but given a priority or weighting so that at least one is 'open-ended'. At a later stage the client must consider all the following functions.

- *Planning*. This involves forecasting his future needs.
- *Organising* — deciding who is to be involved and to what extent delegation of functional responsibilities is to be made. Is there to be a project manager? Or will the architect carry these responsibilities during the initial conceptual design stage? If so, will this continue into the detail design stage? Will it continue beyond?
- *Communicating and motivating*. Is this function to be fulfilled by the client or someone in his organisation, or is a specialist to be appointed, or will the architect or the contractor be given the task? Who will then be responsible for controlling the operation?

Looked at in management terms, it becomes clear that there will also be contractual implications arising from the decisions taken, and indeed it is a lack of clarity and definition during the brief-getting exercise which frequently is at the root of many disputes arising out of construction projects — the price of non-compliance, in Crosby's terms.

Controlling the design
The next stage illustrated by the pyramid is the need to co-ordinate and control four facets — function, aesthetic, cost and time — when the responsibility for the design development is delegated to the various specialist designers.

It is virtually impossible to delegate the overall initial conceptual design to other than one mind. The architect creates at least in outline the client's total spatial objective — the whole building — having first analysed all the client's functional requirements and also considered the constraints imposed by the site, the cost and the time limits. In the process he will have created a three-dimensional solution which will have its own aesthetic, and it is to be hoped that this will balance or fulfil the client's objectives in this area also.

This outline design, or in management terms a feasibility study, should then be considered by the client in relation to all the objectives stated in the brief, at which point many previously unstated objectives or desirable criteria often emerge. These should at least be given a hierarchy of desirability before the architect or overall designer restates them in the form of detail objectives to the other members of the design team, who will then be responsible for their detail development.

Controlling the cost

The management role of the quantity surveyor is frequently misunderstood. At this stage his role is a 'staff' one rather like that of a budgetary controller, and at no stage during the design should he have any executive or 'line' relationship. That is to say, the quantity surveyor measures and assesses the likely cost implications of each part of the design and reports these back to the designer, particularly where they show a deviation from the initial cost plan or design budget. The specialist designer must then take whatever corrective action is open to him, or the overall design co-ordinator (the architect?) must amend the overall design or seek a fresh mandate from the client.

The tendering process

Agreement having been reached on the balance between function, aesthetic, cost and time at the completion of the detail design, the necessary information on which to obtain a tender can be prepared. This can then be put in front of one or more contractors in a suitable form so that they can bid, and put their 'offers' before the client. If any offer is acceptable in terms of time and cost, the tender can be accepted and an agreement then drawn up for the implementation of the client's defined objectives on site. However, this process of tendering should not be the first time that there has been a conscious recognition of the need to clarify the objectives as they have evolved in the design process.

Obtain, assess, accept

Inherent in the taking of any instructions are three processes — obtain, assess, accept; these should have been applied at each stage of the sequential design process (see Table 9).

Once again the need emerges for good management

Table 9. **Process of accepting instructions**

Stage in RIBA plan of work	Progress stage	Task	Responsibility
A	Primary brief	Obtain	Architect
A	Feasibility	Assess	Architect and building owner
A	Initial report	Accept	Building owner
B, D	Secondary brief	Obtain	Architect and design engineers
C, D	Outline proposals	Assess	Architect and design team
D, E, F	Scheme design	Accept	Building owner
E, F, G	Detailed design/ specification	Execute	Design team
H, I	Tender documents	Assess; prepare bid	Contractor
—	Bid or offer	Assess/ accept	Building owner
K	Construction on site	Implement	All contractors

processes to balance (client) requirements against (construction) resources. This need must be given full consideration and total objectives must be identified at the brief-getting stage when the fulcrum is high (in Fig. 6) and the balance can be seen to be precarious. This is a matter of some importance to both client and designer, and perhaps too the professional indemnity underwriters!

The assessments made at each of the stages should enable the process to proceed. If a new contract is to be brought into effect with another party for a successive stage, a pre-appointment audit should be made to ensure that the new organisation is compatible with the existing project team. The existence of a quality management system appropriate to the scale of the organisation's involvement will be evidence of this, but of equal importance may be the 'people-factors' and personalities (see Fig. 16).

Management and the building agreement

When a building project is sent out to tender and a contract is entered into under JCT 1980, or similar, the basic assumption is that all of the many decisions have been made which are necessary to give effect to the building solution of the client's requirements. The contractor has been offered the opportunity to quote his price and time conditions for executing the work in a tender, and the contract is entered into on this basis. All will recognise, as do the normal forms of contract, that variations will occur.

The legal principle which is enshrined in the agreement is that it is a fair contract freely entered into between both parties — the building owner (the 'employer') and the contractor. The agreement therefore assumes that, at the time of signing, the objectives of the various parties involved are as one, or at least that a fair balance has been achieved. The employer has defined (or had defined for him by his design team) the size, shape and quality that he requires, and (by accepting the tender) what he is prepared to pay for it. The contractor knows what is expected of him and when, and what profit he is aiming to make from the project for the effort and resources he has planned to commit to his tasks. The process of pre-contract auditing, preferably as a pre-selection process, will have given both parties the opportunity to assess the contractor's suitability through the evidence of his total quality management system.

But new situations develop, both because unforeseen events happen and because the client, building only once, can change his mind as he has second thoughts or sees new opportunities open before him. The contract provides for amendments to be made and prescribes procedures for such variation orders to be issued so that they do not vitiate the contract. It has also been customary to provide that any differences that result between the employer and the contractor be resolved by the quasi-arbitration of the architect or engineer. But the law, as evidenced by recent judgements in the courts, regards the architect or engineer as the client's agent.

It is also frequently the case that the instruction needed and so provided through the medium of the variation order is necessary as a result of the fault or omission of the architect or engineer, or is the outcome of communication problems within the design team, which may in their turn be the result of communication difficulties between the client organisation and the design team.

All of this is a complex process well removed from section 4.4.6 of ISO 9001. It is also an impossibility for normal third-party certification. However, it is a matter that must be dealt with by the project total quality management system and within the contract documentation.

This situation of conflict provides an illustration of the difficulties in what is known as sideways management.

The four directions of management application

There are four directions in which management flows. The most easily recognised is downwards from superior to subordinate — director to directed, boss to worker. The second is upwards, and every building professional knows that if he cannot 'manage' his client then he will not be able to control effectively the third direction for his management activity — inwards — himself and his own operating work functions. Every manager has two kinds of task: 'operating work' — that which only he can do, such as thinking, drawing, calculating and writing reports — and 'managing work' — getting work done through others.

The fourth direction, that of sideways management, is readily recognised in large organisations as one fraught with difficulty. This is where colleagues of equal status and authority cannot agree on matters where their authority or influence overlap, or where their own planned objectives or procedures are in conflict with those of some other department. Normal management structures resolve this situation by reference upwards.

The new concepts of total quality management aim to solve more problems of this kind without the reference up, and by shallower management pyramids, quality circles and the like. In construction the principles and philosophy are the same but the contractual nexus, the one-off situation and the culture differences and relationships make the task more difficult.

Conflicting requirements

At the design stage of a building, an overlap or conflict in responsibilities can frequently occur. The heating engineer wants to site a duct at a particular position in relation to the heating chamber; the structural engineer needs a column in the very same place, or perhaps wants a beam to pass through it; while the architect, for visual and access reasons, wishes to have a glass door. All three designers have equal status — or at least their technical priorities seem to them of equal priority — particularly if all are working to a tight cost plan, and amendments to their designs will cost more. Normally the architect or engineer as the co-ordinating designer adopts the role of supervisor/manager/judge and makes the decision between conflicting priorities. If, however, the client has separate contracts with each member of the design team, a project manager is needed to make the ruling.

The form of building agreement establishes an equality of relationship between the parties at the start of the project on the site, but then different conflicts of interest develop.

Third-party adjudication as a quality process

In a building project, as indicated above, generally the organisation and relationship patterns stimulate the likelihood of dispute; for example, when situations arise such as the discovery of poor footings where rock was expected, or when unexpected rock is found to be present where a void is to be created, not to mention the vagaries of the weather and actions of outside authorities!

A ruling is needed between the parties now in a sideways management situation in which they could quickly become locked. The courts have found that a ruling from a project manager is not always a valid concept. The function of project manager is relatively new, and the title and function may have been accorded to someone employed by the client to manage downwards or by the contractor to manage downwards and upwards; but despite the best total quality management practices, in neither case will he have the necessary authority to bind both client and contractor.

In this situation, what is needed is someone whose authority is accepted by the parties, who can redefine or represent their objectives to produce the balanced situation that was present in the first agreement. This will be essentially to achieve the completion of the client's building project as close to function, time and cost as originally stated, with satisfactory achievement also of the project objectives for the contractor — and perhaps also to provide adequate sapiential satisfaction for the professionals from the performance of their design and supervisory tasks!

It is preferable that this authority should be exercised as soon as the difference has arisen. Successful football matches, from the point of view of player and spectator alike, result from immediate and firm decisions from the referee, whether these are to allow the game to continue and the players to find their own level or to penalise or advantage quickly one side or the other.

It follows that adjudication should be available immediately from someone outside the game and the players so far. He must be someone given authority by both sides, and accountable only and equally to those separate parties. He will owe a duty of care to both, and not through an agency from one or the other, with its consequent legal implications.

The British Property Federation in 1983 analysed the management procedures of building projects as it saw them. It set up a multidisciplinary working party, with skills developed in all parts of the industry, which came to such a conclusion after many months of close study, it would seem, of both strategy and tactics.

The working party recommended, first, that better management and more thought at the outset of a project would reduce delays later. Secondly, it recommended that good communications and clear responsibilities within the design and development team could avoid future problems and the resulting unnecessary expense and delay; but, nevertheless, that it was also necessary to introduce the concept and role of an Adjudicator, or, rather, two of them, at different stages, the first with duties related to differences between the design team and the second in relation to works on site.

This process of adjudication is akin to third-party assessment in that it requires access to and judgement of the conduct of the parties within the project quality management system. However, if it is to add value to the process it will require a technical knowledge of the process and of the legal framework through which it is conducted.

The use of an adjudicator who, through his monitoring of the project process, will be immediately available to make a judgement and an interim award on the quality factors of function, cost and time will greatly reduce the costs of poor quality and later dispute resolution. A fuller exploration of this 'third-party certification' is made in chapter 13.

Management and motivation
The human element which is present in construction management situations — Dempsey's 'getting work done through people' — is now considered.

What motivates men? What motivates designers in drawing offices? What motivates men on a construction site? This will depend partly on the conditions under which they work. There will certainly be different attitudes to working in exposed conditions in the summer with the temperature at 30°C and in the winter at -10°C . . . or will there? As far as the worker is concerned there may be no difference! In both situations the attitude could be to get the work done and get to hell out of it — or just to do the latter — until the temperature changes.

The reaction of human beings, according to Maslow, depends fundamentally on the extent of their ascent up their basic hierarchy pyramid of human needs (Fig. 7).

Behaviour and motivation
Behaviour is the individual's total response to all motivating forces. One of these forces is the particular situation at a particular time. However, Maslow suggests that all basic human needs can be expressed in a hierarchy of prepotency, with the appearance of one (superior) need usually resting on the prior satisfaction of a lower or subordinate human need. The five levels in this hierarchy are, in ascending order, physiological, safety and comfort, social, egoistic, and self-realisation (Fig. 7).

The theory postulates that man's animal wants are perpetual, and each drive is related to the state of satisfaction or dissatisfaction with the other drives.

Motivation, however, is human-centred rather than animal-centred and is goal-orientated rather than drive-orientated. All rational human behaviour is caused: we behave as we do because we are responding to forces that have the power to prompt — motivate — us to some manner or form of action.

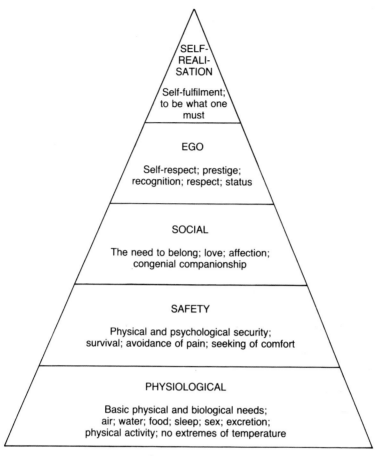

Fig. 7. Maslow's hierarchy of human needs

In a sense, therefore, behaviour *per se* can be considered to be an end result, a response to basic forces.

However, behaviour is actually only an intermediate step in a chain of events. Motivating forces lead to some manner or form of behaviour, and that behaviour must be directed towards some end. That is to say, there must be some reason why we are responding to the motivating force. And that reason could only be to satisfy the force which in the first place motivated us to behave. Consequently, all human beings, whether they do so rationally or irrationally, consciously or subconsciously, behave as they do to satisfy various motivating forces, says Maslow.

The forces that motivate people are legion and vary in

degree, not only from individual to individual but also from time to time. They range from ethereal and psychological to physical, instinctive, and basic physiological forces such as hunger, thirst and avoidance of pain.

Motivation is not synonymous with behaviour. Motivations are only one class of determinants of behaviour. Whereas behaviour is almost always motivated, it is also almost always biologically, culturally and situationally determined as well. We are, in short, the product of our environment.

Undoubtedly, physiological needs are the most prepotent of all needs. In the human being who is missing everything in life in an extreme fashion it is to be expected that the major motivation will be the physiological needs rather than any others. If all needs are unsatisfied, the organism is dominated by the physiological needs; all other needs may become simply non-existent or be pushed into the background, for consciousness is almost completely pre-empted by hunger.

Attempts to measure all of man's goals and desires by his behaviour during extreme physiological deprivation are blind to many things. It is true that man lives by bread alone — when there is no bread. But, when there is plenty of bread, at once other (and higher) needs emerge and these, rather than physiological hungers, dominate. When these in turn are satisfied, again new (and still higher) needs emerge, and so on. That is what is meant by saying that the basic human needs are organised into a hierarchy of relative prepotency.

Thus, gratification becomes as important a concept as deprivation in motivation. It releases the organism from the domination of a more physiological need, permitting the emergence of other, more social, goals.

When physiological needs are relatively well gratified, there emerges a new set of needs, categorised roughly as the safety needs. As in the hungry man, the dominating goal is a strong determinant not only of his current world outlook but also of his philosophy of the future. Practically everything looks less important than safety. A man in this state, if it is extreme enough and chronic enough, may be characterised as living almost for safety alone.

Unlike infants, when adults feel their safety to be threatened they may not be able to see this on the surface. The need for safety is an active and dominant mobiliser of resources only in emergencies, such as war, crime waves, neurosis, and other chronically bad situations.

If both the physiological and the safety needs are fairly well gratified, the needs for love, affection and belongingness will emerge, and the whole cycle will repeat itself with this new centre.

All normal people in our society have a need or desire for a stable, firmly-based, (usually) high evaluation of themselves, for their own self-respect, and for the esteem of others. Self-esteem is soundly based on individuals' own known real capacity and achievement rather than just respect from others. Individuals' needs are, first, the desire for strength, for achievement, for adequacy, for confidence in the face of the world, and for independence and freedom. Second, there is the desire for reputation or prestige (defined as respect or esteem from other people), external recognition or appreciation.

Even if all these needs are satisfied, it may still be expected that a new discontent and restlessness will develop, unless the individual is also doing what he is well fitted for. A musician must make music, an artist paint, a poet write, if he is to be ultimately happy. What a man can be, he must be. This need is called self-actualisation, the desire for self-fulfilment, to become everything that one is capable of becoming.

The specific form that these needs take will vary from person to person. One individual may have the desire to be an ideal mother, another to paint pictures or be an inventor. The drive is not necessarily a creative urge, although in people who have any capacity it will take this form.

The clear emergence of these needs rests on prior satisfaction of the physiological needs for safety, love and esteem. People who are satisfied in these needs are basically satisfied people, and it is from these that the fullest (and healthiest) creativeness may be expected.

The practice of total quality management, by committing all to the recognition of customer satisfaction, and by encouraging shallower management pyramids, team-working, and decision-making at lower levels in the organisations, serves to accelerate the ascent of the individual up the hierarchy pyramid — but in construction always with the potential for conflict.

McGregor's theory X and theory Y

Douglas McGregor takes this behaviour pattern on into what he postulates as theory Y — that people are self-motivated and will respond to what Drucker called 'management by objectives'. In contrast, 'management by control' — theory X — he says, results in people seeking to fulfil their social and self-fulfilment needs away from the job. Theory X, the more conventional management view, is that management is responsible for organising all its resources for economic interests, and this means that people are 'directed' to fit the needs of the organisation. The view is that without this firm direction people would be passive, being by nature idle, lacking in ambition and resistant to change.

Construction workers, perhaps because of their inherent job satisfaction — the carpenter who lavishes his skills on creating mouldings on doors, frames and staircases, the bricklayer whose prowess with an elaborate decorative brick bond will be seen and admired by the generations that pass by his work — all appear to fit the propositions of theory Y better than those of theory X. Even more, many people think that every architect and most designers are all the time engaged on fulfilling their theory Y needs at the top of the hierarchical pyramid!

However, in building construction (as illustrated in Figs 5 and 6) the progress of any project involves many people whose objectives are widely divergent — or perhaps convergent and on a collision course. The behaviour of the parties is the result of factors present in conflict situations. Total quality management practices must aim to reduce these factors, but it will never completely solve the problem of conflict without prevention measures focusing strongly on human behavioural aspects, to change first attitudes and then cultural differences.

Cultural and group influences on behaviour

The way an individual behaves will also depend on his relationship with the other individuals with whom he is in daily contact.

In the 19th century the freedom of choice for the average worker was very limited. Economic freedom existed for very few beyond the employer. In the early days of the Industrial Revolution in the 19th century, the individual could be considered to be hemmed in by the organisations and people around him (Fig. 8(a)).

In the 20th century economic conditions allow far greater freedom of action for the employed individual. The same relationships with family, friends, church, trade unions, the employer and the state exist, but their influence, either constraining or stimulating, has changed, at least as far as

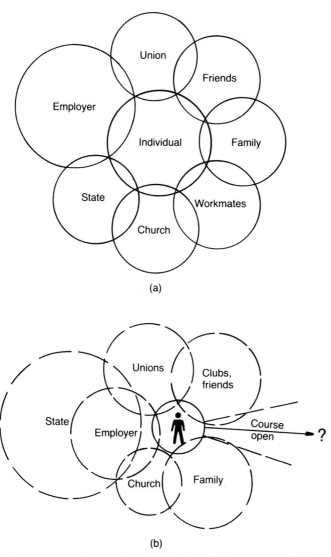

(a)

(b)

Fig. 8. Influences on the worker: (a) in the 19th century; (b) in the 20th century

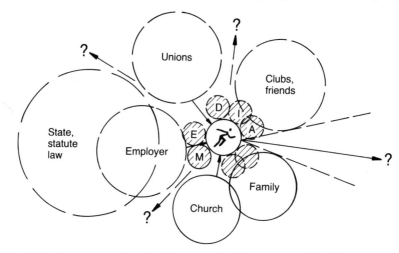

Fig. 9. Influences and attitudes to action in a stress situation

most individuals are concerned, from that experienced by a worker during the early Industrial Revolution (Fig. 8(b)).

When this analysis is taken one stage further, into, say, a dispute situation in the 1990s, the position is different again. The relationships between the individual and the institutions, whether family, friends or authorities, are perhaps less immediate and influencing than they have ever been.

Since the Second World War children have been encouraged by their schools to 'do their own thing', and the economic climate of Britain in the 1980s allowed them to exert their free will to choose, even if the choice has led to thousands of people spending thousands of pounds to become drug addicts, and so be no longer master of their own destinies — and in a horrifying way!

However, the media generally, and television in particular, now have a greater influence on the average person. There is a much more direct, if one-way, communication, with events such as those witnessed daily on television in the home during the miners' strike of 1985. The selection of shots by the programme producers to show horseback charges by the police against 'peaceful' pickets, and the throwing of bricks and flaming torches by these pickets, had more direct influence on a striking miner than the pamphlets or letters sent to him by his union leaders. After all, seeing is believing. The subtleties

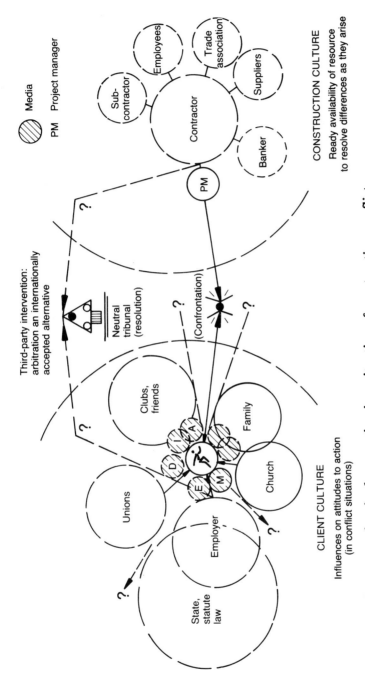

Fig. 10. Influences and attitudes to action in a situation of construction conflict

of selection, first by reporters and then by producers and by programme presenters, are not so apparent but clearly they have an effect on the viewing individual's attitudes and actions in these conflict situations (Fig. 9).

The analysis is now applied to the parties to a construction dispute. It shows the benefits of introducing a neutral adjudication tribunal to bring about a solution to the dispute (Fig. 10).

The tribunal must recognise the management culture of construction and must study, Ishikawa-fashion, the elements that caused the situation — which may develop from materials, machines, methods or money — but above all it must take account of the motivation of men.

What is needed is the giving of authority to a third-party tribunal by all the parties involved, with some predefinition of the tribunal's responsibilities but with the tribunal having clear accountability to all parties for the discharge of these responsibilities.

Application of total quality management and ISO 9000 to the construction project

The differences between the construction project and other processes are now analysed with regard to the application of total quality management practices, and it is shown why third-party certification of the project quality management system (as seen for the product manufacturing processes) is irrelevant and without added value.

Uniqueness of a construction project. A construction project is not even a prototype. It is a single product run. It has a unique production location. Even housing estates with similar house types will have different positions for drainage connections to manholes and main sewers, and for water, electricity and gas service ingress. The ground profiles and possible substructures will be different.

These are all factors which will have influences on detail if not on the basic design. These factors will produce variety and eliminate the potential for any kind of statistical process control. In addition, the construction process will be carried out in the open and different projects will start, finish and be continued in different climatic conditions.

Length of project. Amendments to the project requirement must be permitted both as the design develops and even later during production on site, as the end product has a very long life cycle.

Difficulty in defining quality standards. These will relate not only to the quality of the building but also to the quality of its various parts. The functional requirements and spacial relationships of the building evolve from the design development phase, in which the requirements of services and structure — including their vertical and horizontal implications — interact with aesthetic and spatial requirements. Standards can be defined for materials and manufactured products which are used, but until their location and juxtaposition in a building have been established the interactive effect and standards of control can sometimes not be appreciated.

Uniqueness of people-relationships. Many of the various contractors involved may never have worked with or alongside other firms. Many of the operatives may never have worked with or even met others from firms with specialist input around whom they must work or with whom they must co-operate. Some of the project team will be present throughout the project, others will come and go. Some project teams will come from large company cultures with an engineering background, others will be from a small-company culture with craft skills and background.

Feedback. This is difficult. The construction cycle is long; the feedback cycle is much longer. Sometimes, too, the information coming from feedback will have little or no relevance to other projects in which the firms involved will engage.

The fee scale for the designers, who might benefit most from feedback of success or failure, does not readily permit the cost of the collection of this information. The client, who has depressed the fees and taken the lowest option, does not see how it will benefit him anyway; neither do his bankers or backers.

Difficulties in establishing cost in use. The design criteria for a capital project should be based on the costs in use of the building services, systems and finishes. However, their etablishment and collection has defied many endeavours in the industry since the 1950s when their implication was appreciated.

Consumer conflict. There is frequently conflict between the requirements of the purchaser and those of the product user: the building client, who commissions the project, is generally concerned with different aspects of cost (and their taxation implications) than the lessee or tenant, who by his lease arrangements may be responsible for the cost of maintenance. (Often the members of general public who visit or see the building are the most critical of the project!) Thus, the briefing stage for the design of the project — and for establishing appropriate cost and time elements for the design and production — is most difficult.

Lack of experience of client. The briefing operation is generally further complicated because the client is an amateur in the sense that he rarely is concerned with the commissioning of more than one building and so could consider that there is little incentive to become adequately expert in making what are the top-management decisions for his project. These are the strategic decisions that in corporate structures would always be taken by top management. What is built? Where it is to be built? When it is to be built? To what standards should it be built? (The client has difficulty also in perceiving and defining these.) And — in a people situation the most important — who will be involved in its creation and construction? Arising further from interfaces between these various strategic decisions, there will arise the final difference between construction projects and other processes — the question of the contract.

Nature and form of the building contract. The nature and form of the building contract is vital, and perhaps the primary education that a building client should be given. There are in the UK 94 different standard variations (i.e. printed forms) which have been developed and created by the various groups in the construction industry. Selection of the right form of contract but the wrong structure or design and build process activities can establish impossible functional and legal people-relationships.

Total quality management can do a great deal in the preparation and pre-site stages. It can help develop teamwork across the boundaries of culture and ensure that it takes place.

The parties must above all recognise the client and his contribution as the strategic top manager of the construction process and their aim should be to develop the team approach and put into practice the concept of partnership between client and all contractors.

Internal and external customers

The 'customer' concept in total quality management says that everyone should seek to identify what his customer needs. This must be coupled with the idea that everyone has a customer both outside and inside the organisation.

In manufacturing industry the internal customer is seen as the next group down the line. For example, on a motor-car assembly line the group who place the chassis has for its customer the group that places the engine on it; that group's customer is the group that fit the bonnet and axles; further customers fit the wheels and then the seats and then fix the trim, until at the end of the line the cleaning group is left with little to do before the ultimate external customer proudly drives his finished vehicle away — perfect in all respects. Even more simply, when a boss dictates a letter to his secretary, she is his customer and should be treated as such. Conversely, when the secretary delivers the letter with its enclosures and envelope neatly presented and in the right order, she does so to her customer rather than to her boss.

The implications of this concept in relation to a construction project are vast and dramatic. The last group in the construction of a building are the painters and decorators or floor-layers, generally specialist subcontractors. The total quality management philosophy makes them the internal customer of the general contractor!

The general contractor in his turn is the customer of the quantity surveyor for the bills of quantities. The quantity surveyor is the customer of the architect and various design engineers for the drawings and information on which the bill of quantities is based.

If this is extended further it places the design team as the customer of the project manager, if there is one, or the client! The philosophy behind this concept may be no more than the age-old 'do unto others as you would be done by', but is the culture of the construction industry so different that the message is invalid?

Team-working with the client as part of the team in a genuine partnership to achieve project objectives can and does work, given the right leadership and contractual relations, all part of the project quality management. The better clients, the better professionals and the better contractors know this already, and enjoy repeat and continuing business to their mutual satisfaction. But with the unique project orientation of most construction projects, repeat business is far from the norm, and cultures have developed that require substantial cultural change for client and contractor, design professionals and specialists.

Management audits in construction

Where then does auditing fit in this total quality management construction culture?

The most important audit in construction companies and design firms is the same as in any other organisation. This is a first-party audit, which a supplier does to his own organisation to ensure that the whole operation is functioning effectively and as management sees it should. The quality management awards illustrate this fact.

The concept of first-party comes from the simple buyer-and-seller contractual relationship. The supplier is the first party to the contract and the customer is the second party. Thus a second-party audit is an audit done by a customer on his supplier. The most significant and beneficial second-party audit is that which is done by the would-be customer on his would-be supplier in order to satisfy himself that the supplier has a quality assurance system for the projects or subcomponents that the customer wishes to incorporate in his own manufacturing process, as part of a larger project that will eventually reach an end-user customer.

Sometimes the audit will need to address both design manufacture and supply (ISO 9001). Sometimes the design may be provided by the customer himself, or some other party, and the audit will therefore be of the manufacturing process only (ISO 9002).

In the case of manufacturers of products or components that

Table 10.   The first 17 bodies to be accredited by the NACCB

Lloyds Register QA Ltd (LRQA)
Certification Authority for Reinforcing Steel (CARES)
BSI–QA
British Approvals Service for Electric Cables (BASEC)
Yarsley Quality Assured Firms (YQAF)
Ceramics Industry Certification Service (CICS)
Loss Prevention Certification Board (LPCB)
Bureau Veritas Quality International (BVQI)
Ready-Mixed Concrete (QSRMC)
Associated Short-Circuit Test Authority (ASTA) certification services
SIRA certification services
Construction Quality Assurance (CQA)
Det norske Veritas Quality Assurance Ltd (DnVQA)
Associated Offices Quality Certification Ltd
National Inspection Council Quality Assurance
Engineering Inspection Authorities Board (EIAB)
TRADA Quality Assurance Services Ltd

seek many customers for their standard products, second- and third-party auditing is cost-effective and particularly useful. Thus, having an accredited agency instead of customers to carry out the audit against the appropriate manufacturing standard, and to certify that the supplier has a system which complies with the standard, makes practical and economic sense, provided that the certifying agency is fully aware of the implications of the system within the existing industrial context.

The National Accreditation Council for Certification Bodies (NACCB) was established by the UK government to ensure that no monopoly was created in the certification process but that there was fair competition as well as competent standards, with certification agencies themselves in a competitive market but maintaining good certification practices and procedures. The idea was part of a self-regulating concept for industry, and financial assistance was given to industrial sectors for each to set up its own certification agency that would be experienced and knowledgeable in the particular sector, for example the Certification Agency for Reinforcing Steel (CARES), the Ceramics Industry Certification Service for the ceramic industry and QSRMC for ready-mix concrete. Accreditation of a certification agency by the NACCB also defines the scope of its accreditation (i.e. the area of expertise of its assessors and their assessment systems); all accreditations relate back to ISO

9001 and ISO 9002 or ISO 9003 for the manufacture of products. In construction, the standard of performance required is essentially project-related.

There are now in the UK at least 30 certification agencies. At the time of writing, approximately 20 of these have been accredited by the NACCB and had their scope of competence defined. Table 10 lists the first 17 accredited agencies that carry out third-party audits.

The application and operation of these auditing concepts is described in more detail in part III. Audits — or management-system assessments — are essential components of a total quality system. First-party audits should be conducted not only as a means of ensuring compliance with the management system, but also as the source of benchmarking of best practice. This benchmarking is an important means of gradually, continually and permanently improving performance — even in the construction industry.

Benchmarking for total quality management

As the practice of total quality management has developed in the bigger, the better and the brighter companies across a world spectrum of manufacturing, the concept of benchmarking has grown. Benchmarking is the means by which various aspects of an organisation's performance can be compared against the best practice in other companies. In the 1960s an organisation Interfirm Comparison was set up, with a subscription membership, to provide similar information as part of the productivity movement of the time. Now, as then, by the nature of the construction industry, with its essentially project-based activities in different locations, the concept and principles of benchmarking are difficult to apply, and the lessons to be learnt more difficult to deduce.

To monitor the progress of a development programme for total quality management, it is possible to use 'artificial' benchmarks, based on a table such as that developed initially by Shell Netherlands and subsequently simplified by Polycon for use with smaller companies in the construction industry. An example, scored as for a particular organisation involved in the construction industry, is shown in Table 11. The use of such a table can give a progressive numerical rating while a programme is developing over say, four years, but it is difficult to find realistic areas where assessments can be made on a comparable basis in normal operations.

Financial ratios, such as are normal for stock-exchange analysis, can be extended into turnover per productive man-year or man-day, per salesman-year, per head of office staff, and so on, but because the size and nature of individual projects varies so widely such comparisons, even within a firm, will not be sharp or give incisive guides to operating performance. They may suggest poor strategic management at board or partner level, but will have little benefit at operational level. For example, if an organisation wishes to improve on the number of tenders won in relation to the number of estimates prepared, and the time spent on this (and thus improve the profit margin), making a judgement decision at the commencement not to bid can be more effective than improving the process of estimating!

Areas of management weakness might be indicated if personnel management, for example, were to look at records of absenteeism, lateness or sickness among office and site staff, but few companies in construction will have a sufficient number of people in stable work situations for the statistics to be meaningful or worth collecting.

An overall benchmarking in the 'people' area against assessment for the national Investors in People standard will give a very good idea of the company's progress towards total quality management through training, but these also are almost 'artificial' benchmarks.

An example is taken from a client company, Ardmac, operating as a specialist contractor from several UK bases. Two surveys were taken, one of the company's managers and another of its total workforce. The aims were to establish the views of managers and those within the organisation on the extent to which the company was already committed to the four principles of the Investors in People national standard, and then to establish a benchmark from which improvements in performance could be measured in a year's time. The four principles are as follows.

- *Principle 1 — Commitment.* 'An investor in people makes a public commitment from the top to develop all

Table 11 (below and facing page). Benchmarks of total quality management: scores are included as for a particular organisation at an early stage in a development programme

Points	A. Commitment by highest responsible line manager	B. Quality organisation and structure	C. Quality training	D. Functioning of management improvement teams
10	Quality improvement is normal part of organisation culture; reviewing and auditing also in other 'units'	Quality improvement is part of day-to-day activities and culture of organisation	Everybody trained; quality a natural component of training programmes	Project approach normal part of daily work practice
9	Communicating quality improvement information to customers and to own organisation	Quality improvement is element of new activities; system improvements permanent; cost saving measurable	All staff have mastery of quality techniques	Systematic identification of new projects
8	Quality improvement is integral part of target-setting and also staff appraisals	Quality audits held; project groups functioning well; proposals systematically implemented	Supplementary quality training available	Experience with project approach very positive
7	Quality improvement is agenda item in meetings; giving lectures and publishing on quality	Quality improvement teams functioning well; regular communication concerning improvements	Middle management trained and these managers give training	Half of all personnel have been actively involved in quality projects
6	Stimulating quality improvement in the organisation, giving lectures and being involved in training	First phase project groups completed; implementation plan actively supported	Training in quality techniques, statistical process control, system description etc.; supplementary facilitator training	Positive results of projects become recognised
5	Having taken part in an improvement project	Regular adjustments made to plans; progress monitored	Whole organisation acquainted with the planned quality programme	Implementation of first projects, with system to hold the gains
4	Involvement in progress of project groups and making necessary resources available	Quality plan established; quality information system set up; fully active quality improvement team programme; facilitators appointed	Facilitators and project leaders trained; part of organisation has received quality awareness lectures	A number of project groups functioning
3	Quality goals defined and published	Steering committee meets regularly; a few quality improvement teams functioning; first system description written; quality improvement team plan created	Quality improvement team workshops commenced; training available for facilitators and project leaders	First project progress reports to quality improvement team
2	Chairmanship of the steering committee or quality improvement team	Steering committee convened; first quality improvement team initiated and quality adviser appointed	Quality adviser trained; managers have followed quality managers workshop training programme	Quality improvement team monitors progressing projects
1	Quality principles known; verbal support given but no personal involvement	Quality focal point appointed but not yet an adviser	Quality training programme selected	First project team monitored
0	Obviously no interest	Structure completely lacking	Quality training not started	No quality improvement projects
Score	3	1	0	0

Table 11 — continued

Points	E. Indications of opportunities for improvement	F. Quality assessment ISO 9004—2	G. Realistic customer specifications	H. Training work instructions	
10	Mentality for acceptance of continuous improvement philosophically achieved	Continuous improvement of systems achieved; regular reviews and audits	Realistic specification reviewed and agreed with internal and external customers before tender submitted	—	
9	Customer— supplier relationship focused on improvement	First-party audit AOK	—	—	
8	Awareness of quality costs; these measured where possible; all processes and system descriptions prepared	Quality documentation complete; all staff trained	—	—	
7	Performance indicators introduced for continuous measurement and control	First internal quality audits and reviews in line with norm; first work instructions completed	Responsibilities for all important procedures established and implemented	—	
6	Quality reviews and audits started	All procedures written; first work instructions written	Specifications for all procedures agreed in principle with customer	Training in writing work instructions normal part of job; where necessary, refresher training	5
5	Performance indicators identified	First procedures written following norm and first staff trained	—	Everyone trained; new staff trained as part of normal company training programme	4
4	Project selection criteria identified and first projects selected	Quality handbook outline and programme plan established; staff awareness programme begun	Improved procedures outlined and first discussions on these procedures held with customers	Training programmes developed further; 50% of staff trained	3
3	Quality improvement team has charted the most important work processes	Work procedures and system sketched in following norm guidelines	—	Training programme established and approved; first training sessions complete	2
2	Quality improvement team has identified most important systems and has ranked them	Choice of norm made and decision taken to write work procedures	Most important areas identified for the establishment of customer specifications	Training explained to staff and discussed with management; programme for and description of how to write and implement work instructions	1
1	System identified for which quality improvement team responsible; criteria defined for selecting those of most importance	Aware of quality norm guidelines	Responsibility established for determining customer specifications		
0	No progress	No programme	No progress	No programme	0
	0	1	3	0	

Total 8 (10%)

Table 12 (below and facing page). Benchmarking against Investors in People standard: manager survey

Indicators	Average mark	Consolidated score*
Commitment		
1. Our most senior managers are committed to developing people	B +	24
2. Their commitment has been made known to all our employees	B	22
3. The organisation has defined its broad aims or vision	B	20
4. Employees at all levels are aware of the broad aims or vision of the organisation	B −	19
5. We have a written plan which sets out our business goals and targets	B −	17
6. We have plans that identify broad development needs for the organisation	B −	17
7. These plans also specify how we will identify the training and development needs of individuals and groups	C −	9
8. These plans also specify how we will meet training and development needs	C −	18
9. We have a clear view of how people at all levels in the organisation can contribute to our success	C +	15
10. People at all levels know how they can contribute to our success	C	11
Sub-total	6 Bs 4 Cs	164
Planning		
1. Our plans clearly identify the resources which we will use to meet training and development needs	C	11
2. We regularly review training and development needs, especially when our business objectives change	C	10
3. We have a process for regularly reviewing the individual training and development needs of each of our employees	D	6
4. Starting at the top of our organisation, we have clearly identified who is responsible for developing people	C	13
5. Our managers have the knowledge and skills they need to develop the people who work for them	C +	16
6. We set targets and standards for our development activities	C	13
7. Where external standards exist, notably national vocational qualifications, we have linked our training targets to them	D	7
Sub-total	5 Cs 2 Ds	76

Table 12 — continued

Indicators	Average mark	Consolidated score*
Action		
1. All new employees are introduced effectively to our organisation	C	13
2. All employees in new jobs are given the training and development they need to do their jobs properly	C	10
3. We develop the skills of our existing employees to help achieve our business objectives	B −	17.5
4. Our managers ensure that everyone is aware of the range of development opportunities open to them	C −	1.1
5. Our managers encourage people to help identify their own development needs	C	14
6. Our training and development plans are effectively implemented	C	11
7. Our managers help those who work for them to meet their training and development needs	C	11
Sub-total	1 B 6 Cs	87.5
Evaluation		
1. We evaluate how our investment in people is contributing to our business goals and targets	C +	14
2. We know whether or not our actions to develop people have achieved their objectives	C	11
3. We evaluate the outcomes of our training and development activities at individual, team and organisational level	C	11
4. Top managers understand the broad costs and benefits of their investment in people	B −	17
5. Our attitude towards training and development programmes demonstrates a continuing commitment to developing people	B	18
Sub-total	2 Bs 3 Cs	71

* For all participants: 3 points for every A grade, 2 for every B grade and 1 for every C grade.

Table 13. Summary of results from Table 12

	A	B	C	D	Score
Commitment indicator	A	6	4	—	164
Planning indicator	—	—	5	2	76
Action indicator	—	1	6	—	87.5
Evaluation indicator	—	2	3	—	71
Total	0	9	18	2	398.5

Table 14. Benchmarking against Investors in People standard: employee survey

		Yes	No	Don't know
1.	Have you been told about the company's commitment to training and developing employees?	17	24	1
2.	Could you explain to someone who does not work for the company what the organisation is trying to achieve?	25	12	5
3.	Have you been told how you can help the organisation to succeed?	19	22	1
4.	Do you know who is responsible for your training?	15	23	4
5.	Have you been encouraged to work towards any qualifications relevant to your job?	12	29	1
6.	Were you given any information about the organisation and how it works when you first joined?	24	17	1
7.	Did you get any training to enable you to do your job when you first joined?	17	25	—
8.	Since joining have you received any training to help you do your job better?	18	24	—
9.	Do you know how to find out about the training and development opportunities open to you?	10	32	—
10.	Have you been encouraged to identify any new skills you need to do your job?	17	25	—
11.	Have your identified any?	26	16	—
12.	Has your manager helped you to develop the skills you need to do your job?	18	24	—
13.	Does anyone talk to you before you attend a training programme about what you should try to get out of it?	4	37	1
14.	After you have attended a training programme, has anyone discussed with you what you got out of it?	7	34	2
15.	After you have attended a training programme, does anyone take steps to find out whether your skills have improved?	4	33	5
16.	After you have attended a training programme, does anyone check that you are putting your new skills into practice?	5	32	5
17.	Have you been told what is happening in training and development in the company?	7	33	2
	Total	245	442	28

employees to achieve its business objectives'.

- *Principle 2 — Planning.* 'An investor in people regularly reviews the training and development needs of all employees'.
- *Principle 3 — Action.* 'An investor in people takes action to train and develop individuals on recruitment and throughout their employment'.
- *Principle 4 — Evaluation.* 'An investor in people evaluates the investments in training and development to assess achievement and improve future effectiveness'.

The manager survey (Tables 12 and 13) covered nine of the company's directors and managers, plus the consultants involved. Over the 29 indicators it gave an average score of 13.74 out of a possible 87, or 15.79% of the maximum.

Few differences were observed between the different regions and head office, but the general view taken by the company's managers gave a lower score than the consultants' opinion. This may be because the consultants were more aware of matters that were in hand than the managers completing the survey!

The company set a target to improve its score from 15.79% to 45% of the maximum (virtually a 300% improvement) in the first year of the total quality management programme.

The employee survey recorded the response of 42 employees (Table 14). These employees represented a total of 1828 months' employment — an average of 3 years 8 months: 66% were happy with their progress since joining the company, but between them they had spent only 56 days away from their job on training.

The concept of benchmarking against best practice is an excellent one. However, as yet, beyond the use of artificial benchmarks in the development of total quality management (as illustrated in Table 11) and in benchmarking against the Investors in People standard (as illustrated in Tables 12 and 13), the author regrettably cannot see how the principle can be applied with benefit to most companies operating in the construction industry.

3 Contractual options and integration with quality systems

This chapter examines what is perhaps the principal way in which quality management systems in construction differ from the systems of ISO 9000 related to manufacturing industry and repetitive processes. The reasons for the differences in the contractual situations are explored and two specific applications of the total quality management philosophy to the building or construction project are proposed: the first application is second-party audit on a pre-contract but project-specific basis for all organisations with a major involvement in the project; the second is adjudication by third-party assessment to make an interim award on claims made during the progress of the works to maintain the team spirit without which quality will not be achieved.

THE BACKGROUND for the ways in which quality systems in the construction industry differ from those envisaged in the ISO 9000 series is referred to in chapter 2.

Also, questions are raised in chapter 2 concerning the relevance of third-party certification for individual construction firms. In manufacturing industry, certification is only the first stage of a two-stage approach: the significant stage from the customer's point of view is the second — product testing prior to the issue of an Agrément certificate or kitemark relating to the product's quality performance. The first stage certifies that the product tested in the second stage comes from a production line which can consistently produce all the other products to the same standard. Such assurance is essential to the customer, who will never receive the actual product or sample which was satisfactorily tested.

In practice, it is impossible to provide this second-stage certification in construction, due to the one-off nature of each project and of the 'team' involved in its design, manufacture, erection and completion. It is equally difficult and cost-effectively impossible to issue meaningful third-party certification. The scope of a manufacturing assessment is defined in each of the ISO standards and the assessment is therefore made against the descriptive sections of that standard. In a construction organisation, whether concerned with design, assembly, erection or even commissioning and maintenance, the scope for an assessment will never be that appropriate to any client's specific needs.

A design organisation may have developed its management systems generally and then have subsets for the type of work it is engaged in or principally engaged in. For example, this may be the design of domestic flats or hostels. The next project which the organisation may be considered for could be the design of a hotel or perhaps a theatre. The requirements for the management system within that organisation will need to be established specifically for that building type.

There will then be further variants of timing, availability of resources, location of the site, interface with other design organisations with whom the organisation in question may not have worked previously, interface with contractors so far unknown, and the selection of other specialist contractors — though first the need for their products or assemblies has to be determined.

Any third-party certificate given by an assessing organisation against ISO 9000 would be misleading to a potential client if he assumed that a general third-party certificate, even from an accredited agency, assured him that the certificated organisation had the resource, competence and management system necessary for his project.

Contractual implementations

Apart from these factors, and a unique site (and site conditions), unique relationships have to be established through which the design is developed by a range of separately organised and controlled designers, some professionals and some within commercial firms, and then the building constructed by a range of contractors and subcontractors, so called because their involvement is developed through specific contracts — very different from the simple contract with a single buyer and a single seller envisaged in ISO 9000.

It is the very existence of these various forms of contract that adds a vastly different dimension to quality management in the construction process. Even the selection of the appropriate 'standard form' is a matter of difficulty for the employer

Table 15.   Standard forms of contract for building and civil engineering work

Form of contract	Arbitration (quality cost) provision

Standard Form of Building Contract, 1980 edition, as amended
  (a)   With Quantities
  (b)   Without Quantities
  (c)   With Approximate Quantities
	Article 5 and clause 41

Standard Forms of Employer/Nominated Sub-Contractor Agreement (NSC/2 and NSC/2a), 1980 edition, as amended	Clause 10 (NSC/2); clause 8 (NSC/2a)
JCT Standard Forms of Nominated Sub-Contract NSC/4 and NSC/4a, 1980 edition, as amended	Article 3 and clause 38
JCT Warranty by a Nominated Supplier (TNS.2: Schedule 3 of the JCT Standard Form of Tender by Nominated Supplier), as amended	Clause 4
Standard Form of Building Contract with Contractor's Design, 1981 edition, as amended	Article 5 and clause 39
Intermediate Form of Building Contract for Works of simple content, 1984 edition, as amended	Article 5 and section 9
Standard Form of Sub-Contract Conditions for Sub-Contractors named under the Intermediate Form of Building Contract (NAM/SC), 1984 edition, as amended	Article 4 and clause 35
Standard Form of Management Contract, 1987 edition, as amended	Article 8 and section 9
Standard Form of Works Contract/1 and Works Contract/2, 1987 edition for use with The Standard Form of Management Contract, as amended	Works Contract/1 Works, section 3, article 3; Works Contract/2, section 9
Standard Form of Employer/Works Contractor Agreement (Works Contract/1), 1987 edition, as amended	Clause 7
Agreement for Minor Building Works, 1980 edition, as amended	Article 4 and clause 9
Fixed Form of Prime Cost Contract, 1976 edition	—
ACA/BPF Forms of Building Agreement, 1984	Clauses 25.1−25.8
ICE General Conditions of Contract, 5th edition	Clause 66
GC Wks 1	—

(building owner), and the degree to which he or his advisers are able to define the quality standards should be inter-dependent with the nature of the contract format chosen. In the UK there are currently 94 different standard variations of forms of contract in use. Table 15 lists some of the principal ones and indicates the clauses under which contractual dispute procedures are dealt with.

Total quality management can do much to prevent some of these disputes arising: these are probably the ones that might be considered the costs of poor quality during the briefing and design development stage. However, the nature of construc-tion is that conflicts arise from site activities which will still require third-party intervention, and so contractual provision is made for it. Auditing therefore must have regard to these various contractual situations at the various stages of the overal construction process. These pre-contract auditing situations are illustrated in Fig. 11.

The quality system developed within any firm which contributes to the project design or building process must have regard to the implications of these contract forms on the firm's quality system; and since many of these firms will be involved in several projects at the same time under different contract forms and project conditions, proliferation of procedures — and even systems — can too easily develop.

These variations of contract forms have been developed not as a perversity of the industry or even of the associations representing the differing interests of their members: they are the answer to different problems posed by the fundamental nature of the industry, involved as it is in solving one-off problems, and having as it does definitions of need (and of standard of performance required to meet that need) which are often changing, not only in the design phase but also until the end of the construction phase. This latter characteristic is related to the fact that building owners involved in a single capital project must have the right to change their requirements at any time right up to the eleventh hour — provided that they accept and pay for the real cost of the amendment!

No study or programme to develop quality management systems to achieve a quality end product — the building — can therefore exclude a consideration of the implications of the range of contract forms available and the effect of this on each

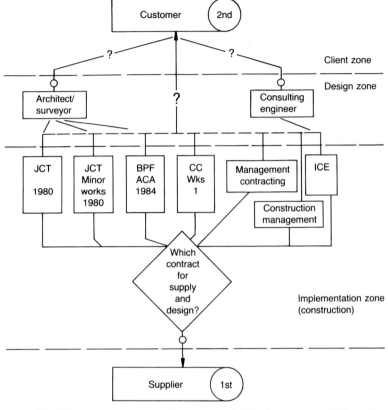

Fig. 11 (a). First-party, second-party and third-party auditing in the construction process: which contract?

firm's quality systems. It is through these forms that the authority, responsibility and accountability of the parties to the contract are legally defined, and the quality system must, as a matter of practice, have regard to these legal implications. Failure to integrate the legal requirements with practical management action over the range of parties and people involved can lead to a situation where disputes (not just complaints) arise, arbitration and litigation ensure, and the quality costs soar out of control.

Specific applications of total quality management
Auditing
Third-party certification of firms engaging in construction projects, as a one-stage assessment applied in a manner similar

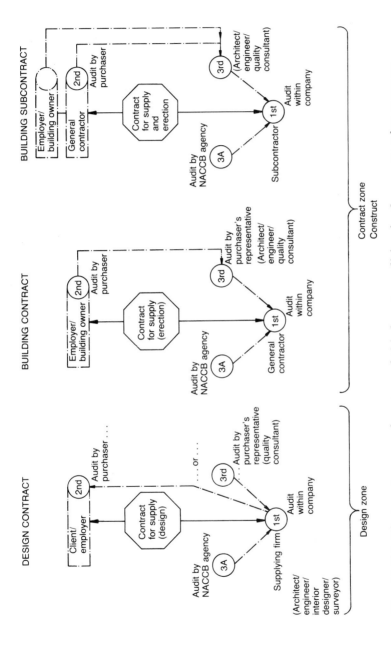

Fig. 11 (b). First-party, second-party and third-party auditing in the construction process: auditing alternatives (NACCB, National Accreditation Council for Certification Bodies)

to that used in other industrial sectors, will not provide customers of construction firms with the same assurance that it gives to customers of other industries. There are dangers to customers of a knowledge-based service in perceiving a third-party certificate in similar terms to those applicable in manufacturing industry and for repetitive services. Therefore, third-party certification is not recommended for professional or construction firms. They should establish their system, and ensure by their own (first-party) audit that it is properly established and maintained. When any potential customer wishes to assure himself, he can do so through second-party *project-specific* audit. This approach will be more reliable, more effective and less expensive than third-party certification, as it will be concerned only with the supplier's suitability and ability to perform on that particular project, at that particular time, through the particular form of contract chosen, and with the particular project team already under contract, or proposed to be so.

Third-party intervention

A further step is possible in project quality assurance, and it is one which will also assist in cost reduction for all parties involved. This is the procedure of contract management adjudication, for which an 'instant umpire' is available to rule both on matters related to the quality of product assembled or erected and on questions related to the project management system, which has been created to run the project in line with the contractual obligations — the legal promises — made by the parties involved. These parties are the client/building owner/customer, the designers — professional or commercial, the main contractor and the many subcontractors. Some of the subcontractors may also manufacture and may have achieved certification both of their quality system to ISO 9000 and of their product tested for a kitemark or Agrément certificate — although neither of these will ensure that their product will not fail when built into or on to the building.

The concept and application of contract management adjudication, a logical outcome of the quality management philosophy related to the construction process and its practical implications, is dealt with in chapter 12.

The implications of these contractual relationships are

highlighted by the standard forms (Table 15) which exist and under any one of which a project team can be assembled and legal promises made.

Frequently, too, individual variations or amendments are made to these forms by the client or his agent. Sometimes these inadvertently go to the root of the contract and further disturb the relationships within the project team, from which the project quality system and plan stem.

Alongside each form is a note of the provision for arbitration, and in the case of the ACA/BPF form for adjudication or arbitration. There is also provision for limited adjudication in the Nominated Sub-Contractor and Standard Form of Sub-Contractor conditions.

Although there are provisions for arbitration (or possibly adjudication) in the standard forms of contract, provision should be made in the project quality management plan, developed by or for the employer, for contract management adjudication. This should be repeated in each contract tier, to designers, to main contractors and to subcontractors. Such arrangements are the final steps in the project quality system.

Such provision will do much to prevent disputes arising from conflicts of interest. (In practice the forms of contract often sharpen such conflicts of interest, whereas they should reflect the mutuality of interest between all members of the project team.)

The project management plan — provided that it has regard to these contractual matters — will then better maintain the active support and participation of firms in all areas of the project organisation and so bring about significant and substantial reduction in quality costs. This will give practical effect to the concepts of teamwork and 'everyone has a customer, internal or external', around which the philosophy and principles of total quality management have been developed for other organisational and industrial sectors.

As for contractors and subcontractors, their tasks and quality systems will also need to have regard to the degree of their design input: this will vary within the contract formats, so affecting their quality management plans (Fig. 12). The degree to which ISO 9001 or 9002 applies to their operations will also depend upon the extent to which they manufacture components or assemblies on a repetitive basis. If there is

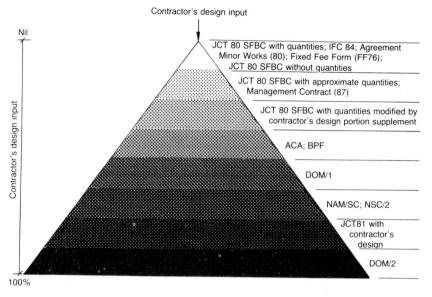

Fig. 12. Contractor's design liability and quality: this diagram is not intended to show the intended extent of design input but rather the likely extent in practice (from Chartered Institute of Building Technical and Information Service, 1990, No. 118)

substantial repetition in the process, third-party certification of the manufacturing element of the process by an accredited agency will be meaningful and may provide added value to the whole process and so to supplier and customer.

But for the firm, its first-party auditors will be looking at everything that must be and is being done to discharge its contractual obligations and quality improved performance!

For repetitive processes, the ISO standard should be the basis of the system and its auditing and assessment. For unique projects, total quality management and the project's specific requirements and standards should be the basis of assessment and auditing.

In construction, this will need a much greater knowledge of the project and its framework and organisation than of the ISO 9000 standard than is practical for third-party certification agencies. However, project auditing has an important part to play in improving the performance and cost-effectiveness of the end product, a quality building.

4 General application of total quality management to construction projects

The purpose of all quality management systems is to organise all the resources and efforts of the supplier of a product or service to the needs of his many customers in his market-place. In construction, the industry with its many separate suppliers must always focus on the particular needs of a single client to provide a single product, the project, for which the client must establish an organisation framework and project quality system (for nobody else can). This chapter takes an initial look at the special features this produces in total quality management of the project.

AS SHOWN in chapter 3, the nature of the contracts between the parties plays a dominant part in the quality system required for a construction project. The quality standards required from the project and the responsibility for achieving them must therefore be specified in the project documents — the plans, specifications, schedules, bill of quantities and so on — and the addition of references to ISO 9000 are likely to be more confusing than helpful.

This has been well recognised by the quality assurance working commission of the International Council for Building Research Studies and Documentation (CIB W.88), which has developed a guide to the application of total quality management in firms of professional consultants, and key elements for quality systems in builder contracting firms. The relationship between this and the ISO 9000 series and its European and British counterparts is shown in Fig. 13. The ways in which the ISO 9000 series needs to be further developed to cover effectively the very different requirements for building are identified in Table 16, which shows the adequacy of the various sections of the ISO 9000 series from the standpoint of the firms involved in building design and construction. The table is based on work which has been done by the Danish civil engineers and members of commission W.88.

Table 16. Relevance of sectional relationships of ISO 9000 series to construction (see notes on facing page)

BS 5750 Part 8	Part 1	Part 2	Construction industry function/processes	Architect	GP surveyor	Quantity surveyor	Building surveyor	Structural/civil engineer	Building services/M&E engineer	Building contractor	Craft subcontractor	Manufacturing subcontractor
5.2.2	4.1	4.1	Quality policy and management organisation	•	•	•	•	•	•	•	•	•
5.4	4.2	4.2	Quality system	•	•	•	•	•	•	•	•	•
—	4.3	4.3	*Contract review	•	•	•	•	•	•	•	•	•
6.2	4.4	—	Design control	•	X	X	•	•	•	C	—	C
5.4.3.2	4.5	4.4	*Document control	•	X	•	•	•	•	C	—	C
6.2.4.3	4.6	4.5	Purchasing	C	C	C	C	C	C	X	X	X
—	4.7	4.6	Purchaser supplied product	X	X	X	C	C	X	X	X	X
—	4.8	4.7	*Product identification	X	X	X	X	X	X	•	X	•
6.2	4.9	4.8	Process control	X	X	X	X	X	C	C	C	C
—	4.10	4.9	*Inspection and testing	X	X	X	X	C	X	•	C	•
6.3.6	4.11	4.10	Inspection/measuring/test equipment	X	•	X	X	X	•	•	C	•
6.3.4	4.12	4.11	Inspection and test status	P	P	X	X	P	P	•	C	P
6.3.5.2	4.13	4.12	Control of non-conforming product	P	P	P	P	P	P	•	P	P
6.3.5.2	4.14	4.13	Corrective action	P	X	P	P	P	P	•	P	P
6.2.4.6	4.15	4.14	Handling	X	•	X	X	X	X	•	P	•
6.4.1	4.16	4.15	Quality records	P	P	•	•	•	•	•	•	•
5.4.4	4.17	4.16	Internal quality audits	•	P	P	P	P	P	•	—	•
5.3.2	4.18	4.17	Training	X	•	•	•	•	•	•	•	•
—	4.19	—	Servicing	X	X	X	X	X	X	•	X	•
6.4.3	4.20	4.18	Statistical techniques	X	X	X	X	X	X	X	X	X

* needs more structuring, or restructuring; • adequately addressed; X not so relevant; P project requirements need expressing differently; C the contracts between customer and suppliers at the different stages should specify requirements, not the standard.

ISO 9002 illustrates by a quality loop the customer-producer quality relationship which underlies ISO 9001 – 9003. This is compared in Fig. 14 with the construction procurement loop developed by the Construction Industry Research and Information Association.

The construction procurement loop is a simplification of the design-and-build process illustrated in Fig. 15. Shown against the project cycle in this figure is the considerable work done in the 1960s by the Royal Institute of British Architects in their plan of work and that done in the 1980s by the British Property Federation. The figure illustrates, as a simplification of both quality management systems, the division of responsibility between designers, architects and engineers, contractors and subcontractors generally, each with a separate contract with a building sponsor and therefore none with an overall contractual interest in effective total quality management of the project.

The Drucker-inspired management structure of planning, organising, leading, motivating and control (Fig. 16) has in

Notes to Table 16
1. The principal context for quality in building design and construction is on the one-off project, contributed to by many firms, sometimes under separate contracts or subcontracts. The product only comes into being at the end of a complex series of processes. The customer himself is much involved but through contractual rather than management control; thus in the case of contract review, document control, product identification, and inspection and testing (Table 16), project procedures and a project manual — the requirements for which should be specified in the contract — must always be more relevant than the ISO standards.
2. As noted, the letter C in Table 16 indicates that contracts specify the requirements. Control is not in the supplier's hands and it is not the standard: the relevant criteria are the client's specific requirements stated in the contract.
3. The letter X is used in Table 16 where it is not practicable or sensible to deal with the one-off cases in the way described in or envisaged by the standard.
4. In each standard, section 1.1 (Scope) states that the requirements of the standard are aimed primarily at detecting non-conformity and preventing its occurrence: the supplier is required to demonstrate his ability to control the processes that determine the acceptability of the product. In the one-off building project, for several reasons, not the least the differences brought about by the contemporaneous forms of contract, the requirements of ISO 9001 and ISO 9002 will *not* provide this assurance.
5. For the design processes, a most important requirement is the provision of an adequate information database or library.
6. For both design and build, a resource planning system to ensure that human resources with appropriate experience will be available for the project at the time required is an essential requirement for project quality and performance.

construction to be applied to the project and clearly this must start at the beginning of the project cycle, as the earlier the price of quality is paid the greater will be its effect.

The present demand by building sponsors to depress fees for the early development of design and planning stages will

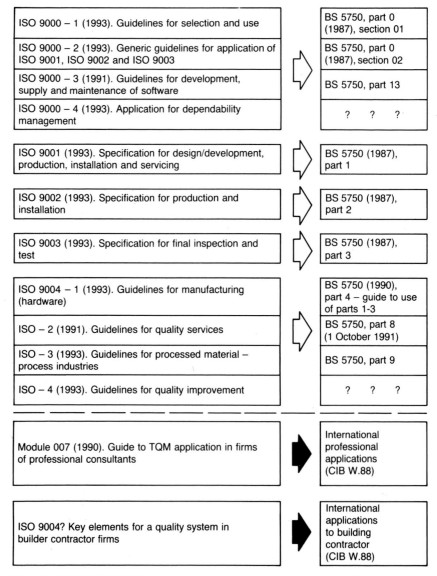

Fig. 13. The ISO 9000 series and its relationship with British standards and the construction process: the ISO 9000 series is identical with the European EN 29 000 series

certainly be counter-productive as regards the costs and performance on a project. Notwithstanding the difficulties of effecting regular working with only a limited range of suppliers, negotiated bids where quality is seen at least in perspective with reputation and price does produce a better

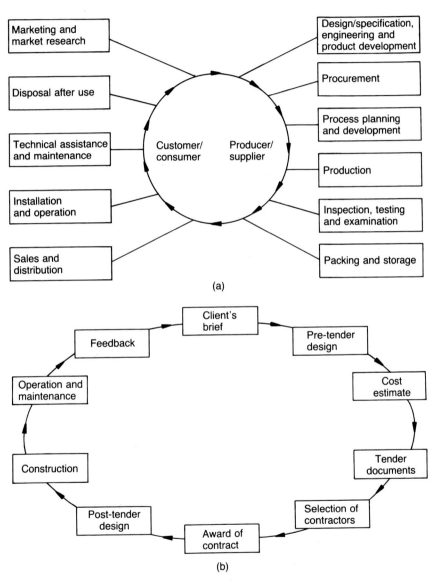

(a)

(b)

Fig. 14. (a) The quality loop for manufacturing (from **BS 5750 part 0**); (b) the procurement loop for construction (**Construction Industry Research and Information Association**, *Quality assurance in construction*, **SP55, 1988**)

end result — when all are unified through an effective project management system after pre-contract audit.

For the design team, innovative and interactive processes within the team and with the client involves processes where knowledge and experience have far greater importance than regulatory systems. Many of the attitudes that the philosophy of total quality management advances have always been implicit in the professionals' ethos of customer service. Over systematisation and excessive control will be ignored or can be demotivating to such individuals — generally at a higher level in Maslow's hierarchy! However, this is not to say that their performance should not be monitored at key stages. Indeed, clear documentation at key control points is essential to effective progress and also to prevent contractual disputes from occurring later.

Both the Royal Institute of British Architects plan of work

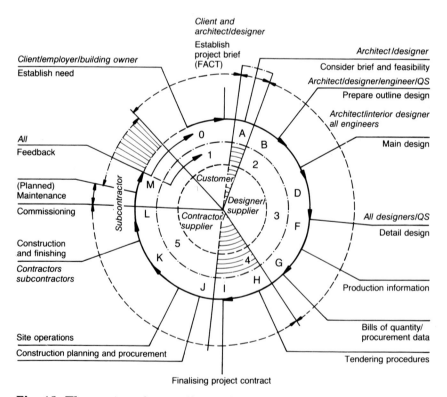

Fig. 15. The construction quality cycle: letters A to O relate to the Royal Institute of British Architects plan of work stages; numbers 1 to 5 relate to the British Property Federation system stages

and the British Property Federation system are far better quality plans and guidelines for total quality management on projects than can be deduced from the present work of the ISO 9000 series — notwithstanding the considerable guidance given in ISO 9004−2, *Quality management and quality system elements: guidelines for services*. (This latter document, excellent as it is, emphasises the implications of quality for repetitive services to customers generally rather than the particular and unique specific requirements for a single project client.)

Total quality management plans for construction should

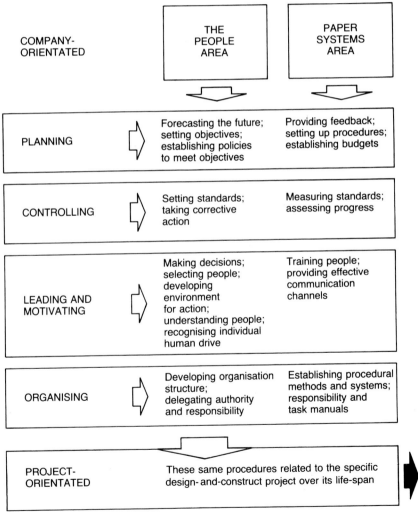

	THE PEOPLE AREA	PAPER SYSTEMS AREA
COMPANY-ORIENTATED		
PLANNING	Forecasting the future; setting objectives; establishing policies to meet objectives	Providing feedback; setting up procedures; establishing budgets
CONTROLLING	Setting standards; taking corrective action	Measuring standards; assessing progress
LEADING AND MOTIVATING	Making decisions; selecting people; developing environment for action; understanding people; recognising individual human drive	Training people; providing effective communication channels
ORGANISING	Developing organisation structure; delegating authority and responsibility	Establishing procedural methods and systems; responsibility and task manuals
PROJECT-ORIENTATED	These same procedures related to the specific design- and-construct project over its life-span	

Fig. 16. **Project management process: people systems and paper systems**

Table 17.   Key elements for a quality system in building contracting firms (from CIB W.88) and corresponding clauses in ISO 9004−2

	Clause in ISO 9004−2
1. *Scope of project resource.* A clear understanding by both the firm and its potential customers of the qualifications and expertise of the firm and the scale of work they can realistically undertake	4.1 5.5.1
2. *Resource assessment.* A systematic method to evaluate the resource requirements of a job against the capacity of the firm, before tender or acceptance	5.3.1
3. *Client brief.* A systematic method to establish the customer's requirements in terms of time, cost and quality; and to identify any consequent need for supplementary specialist subcontractors	5.5.2 6.1.3
4. *Specialist subcontractor integration.* A systematic method to select supplementary specialist subcontractors on the basis of their expertise; the scale of work they can undertake; their quality systems; and their track record; to ensure that the contribution they make is as vigorously controlled as that of the leading firm; that any design contribution is integrated into the main design at an appropriate time and is well controlled	6.2.4.3
5. *Project management plan.* A quality plan relating to the specific requirements of the project: this should include management organisation; quality control; and periodic reviews, involving the customer, to ensure that requirements are being met in terms of time, cost and quality	6.2.4.1
6. *Operational information and method statement.* Methods to ensure that site management and workers have information needed for construction easily available to them and understand the construction procedure required. This includes both directly employed and subcontracted employees	6.2.4.2
7. *Feedback of system performance.* A method to record problems in the operation of the quality system, and to use this feedback to improve the system	6.4.1 6.4.2
8. *Co-ordinated project information.* A method for organising project information which makes it convenient for the various uses to which it will be put. The UK has, for example, recently begun to adopt a method based on commonly subcontracted packages of work. It has also developed over many years an approach to co-ordinating the work of architects, engineers and quantity surveyor, so that all details relating to the project are in a co-ordinated set of documents, the absence of which is frequently the cause of dispute and often of excessive cost in dispute resolution. The technique is known as co-ordinated project information (CPI)	6.2.1

therefore start with the needs of firms providing construction processes. This is well ilustrated in Table 17 for building contractors, developed from the work of CIB commission W.88.

Cross-reference to ISO standards is a valuable final check, as is shown by Table 16, but slavish adherence to systems developed for other situations will not add cost-effective quality to the building project. What it certainly will do is heighten the need for the dispute-resolution process as developed in chapter 13.

5 Total quality management: its philosophy and principles for the building client at the project inception stage

This chapter explains some of the difficulties that unique construction projects develop, and why the building client must be involved in the application of total quality management if the building process is to be successful. As 'top management', the client must delegate, but within a framework of knowledge that will embrace pre-contract auditing of those firms with which he must also interface at all stages and with whose advice he must continue to make the key strategic decisions.

THE DESIGN and construction of a building or civil engineering project is, worldwide, one of the most complex and difficult industrial undertakings. It requires management skills of a high order and is frequently undertaken by firms with little or no formal training in management. To complete the structure fit for its purpose, right first time, and to predetermined quality standards — which must therefore embrace the client's functional, aesthetic, cost and time requirements, around which the project quality plan must be framed — requires not only planning, organising, budgeting, controlling and careful adherence to an overall project management system, but also a high level of understanding of human motivation and behaviour within prescribed contract conditions.

It also requires recognition that quality management systems are never fully within the control of any one organisation involved in the project's contracts. For, in the UK as well as in most other countries, other organisations and individuals can, and do, exercise authority and control over the project, but without accountability to the client — the building team's customer — for the overall performance of his requirements.

This control may affect the appearance, size and general layout of the project through the regulations for town planning, and fire and means of escape; it may affect the combination of materials through building regulations; and it may affect the method of building through the regulations for health and safety, traffic, and noise control. In many cases the control is exercised after the event, and, in some cases the nature of the particular requirements of these authorities cannot be established beforehand.

The principles of quality management developed within the context of a simple buyer and seller relationship — the first and second parties to the contract — need further development for the complex nexus of contractual relationships that must be organised to meet the unique requirements of a one-off project carried out on a unique site at a single point in time through a unique combination of corporate and human relationships.

Quality costs

Experience has shown that the technical failures of product or material are minute compared with the enormous dissatisfactions that arise through failure to meet cost and time targets or even aesthetic expectations, which are all part of the make-up of a quality project. Many of these failures stem from the failure to recognise the extent to which the building client should participate not only in fully exploring and analysing with the designer his full requirements in the early stages at the inception of the project, but also, then and subsequently, in making a full case study of the ability, and suitability of the resources of, those organisations through which his requirements will be fulfilled.

To recognise the importance of the selection process in developing the human team for the project, the client will need tools, and training in their use, to ensure as far as possible that his particular project will achieve success.

Project quality assurance is the aim. The objective is customer satisfaction through a project fit for its purpose, completed on time and at optimum cost, with ongoing job satisfaction to all those involved in its creation.

Role of the client/customer

Every building client must recognise that, in construction, perhaps even more so than with any service industry, the service he gets will depend in part on his own involvement in, and contact with, the supplier of the service.

However much he delegates — and certainly the technical aspects of design and the bulk of the physical aspects of construction will be delegated — he cannot delegate the ultimate responsibility for all decision-making related to the project. He must retain the decisions for what he builds, where and when he builds it, to what standard or level of performance, and within what cost and time limits, and finally, and most important of all, whom he selects to implement his project.

This is a main thrust of this book, and this chapter aims particularly to inform building clients of the minimum extent of the role they must play in achieving the success of their own projects.

This book seeks also to encourage the client's further participation by study of the other chapters in part II to the extent necessary for his understanding of the role of each of the other members of the building team in producing the cost-effective quality building. Some of the necessary tools are provided in part III.

In the context of construction, cost-effectiveness must be judged (unless the brief specifies otherwise) by the ultimate quality and costs-in-use of the building over its useful life-span — not by the cost of the planning, early design and organisation of the project. In short, economy in design and project management fees are a false economy.

The earlier the cost of obtaining quality is built into the design and management process, the greater will be its effect. Thus, the sensible client will seek a design team and project management resource that have established their own management systems and can illustrate them with hard evidence and probably reputation. He should then be prepared to invest more — not less — than the standard fees to carry out the fuller and more effective investigation into construction and construction resources that will produce a lower-priced but better-value building from a better-motivated construction team, who will not have had to price unnecessary risk into the project.

Elimination of unnecessary risk is vital in a one-off project. Quality management and total quality management systems aim to completely achieve this through effective second-party auditing, first at the pre-contract stage and then by monitoring the project management system as progress takes place.

Immediate action

If a client wishes to delegate responsibility as far as sensible for his project, he should appoint an experienced construction-industry auditor to report to him on the suitability of likely members of the design team, and on a quality management system suitable for the project. If, however, the client intends to be further involved in the project — and this I recommend — continued study of this book will help him to carry out audits on prospective suppliers of construction services before any are appointed.

A pre-appointment audit should certainly be carried out on any firm which the client considers employing in order to help establish his brief and to assist with the preparation of an outline design.

Whether such a firm (or individual) is appointed as 'client representative' or 'project manager', it will be essential to establish that it will be able, through its knowledge, experience and total quality management system, to then carry out, through similar pre-assignment auditing procedures, the selection of other potential members of the project team.

In chapters 10–12 these auditing and monitoring procedures are dealt with in more detail, and appendix 3 provides a complete document for the initial approach by an intending building client to this all-important client aid.

The continuation of the second-party auditing procedure (or auditing by an experienced construction third-party assessor on the client's behalf) will incur costs, but will certainly be more relevant to the project than one related to ISO 9001 or ISO 9002 at any random time, and will pay off in the quality of the work and its cost, whether a contract is finally obtained by competitive or negotiated tenders.

An important element in pre-appointment audit of the project manager or design team will be, as it always has been, a personal compatibility between the client — together with any directly involved members of the client team — and the individual or firm to be appointed. It is at this key stage of project inception that there should be a recognition of the human elements involved in cross-culture relationships. The pre-appointment project-specific audit should enable a subjective assessment of these social and cultural factors, as well as of the technical knowledge and capabilities of those involved, to be made.

Team-building and the harmonisation of objectives towards a common goal of a quality building begins, or can fall, at this first hurdle. It is the client who must jump it!

6 Total quality management and the design team through the project design phase

This chapter explains the importance of the design or project management professional educating his clients on the implications of good management for the overall project process and, in particular, on the need to carry out project-specific audits of the organisations that will later make up the quality team. This additional service will add value to the overall process, but should not be at the design professional's expense.

CHAPTER 5 shows how important it is to begin the harmonisation of objectives between the client and those who will be involved in design and construction of complex construction projects. As the design professional will know, this requires management skills of a high order, but all too frequently it is undertaken by firms with little or no formal training in management. The task of harmonisation in the management of the design process must bridge the gap of cultures as well as technical and contractual procedures, a process that will then need to be repeated when the further and socially different cultures of building contractors are involved in the implementation stage on site.

The project quality management system also requires recognition that quality management systems are never fully within the control of any one organisation involved in the project's contracts: in the UK, as well as in most other countries, other organisations and individuals can, and do, exercise authority and control over the project, but without accountability to the client, who is the building team's customer, for the overall performance of his requirements.

Members of the design team will know that this control may affect the appearance, size and general layout of the project through town planning, fire and means of escape regulations, the combination of materials through building regulations, and the method of building through health and safety, traffic and noise control regulations. In many cases this control is exercised after the event, and, in some cases the nature of these authorities' particular requirements cannot be established beforehand.

Building clients, being for the most part building 'amateurs', do not know or understand the implications of all this and rely on their professional advisors to advise them in these matters. The attention now addressed to quality assurance in the UK tends to suggest that all a would-be building client needs to do is ask for a third-party certificate that the firm has a quality assurance system in accordance with ISO 9000 (BS 5750) and all these problems will somehow magically disappear. Architects, engineers and surveyors will know that nothing is further from the truth.

The principles of quality management developed within the context of a management buyer-and-seller relationship — the first and second parties to the contract — need further development for the complex nexus of contractual relationships that must be organised to meet the unique requirements of a one-off project carried out on a unique site at a single period of time through a unique combination of corporate and human relationships.

Quality costs

The technical failures of product or material are small compared with the enormous client dissatisfactions that arise through failure to meet cost and time targets and unfulfilled human expectations, which are all part of the make-up of quality costs. Many of these failures stem from the failure to recognise the extent to which the building client should participate not only in the early stages of the inception of the project in fully exploring and analysing his full requirements, but also, subsequently, by having a full study made of the capacity, and suitability of the resources of, those organisations through whom his requirements will be fulfilled.

Additional fees for management work at the most effective stage

Relating their client's requirements to the contractual resources to be used to fulfil them offers additional and invaluable opportunities for the design team to play a fuller role in the effective management of the whole project by carrying out thorough pre-contract second-party quality audits on potential contractors. These audits should attract a separate fee, and provide not only a service of real benefit to the client, but also one which helps protect the professional's own position in ensuring that the quality of the design is carried through on the project.

The earlier in the project the price of quality is paid, the greater will be the ultimate quality and cost benefit on the end product — the building. But the price should not be paid by the design teams — unless they themselves have failed to develop and maintain their own appropriate management system.

Project quality assurance is the aim. The objective is customer satisfaction through a project fit for its purpose, completed on time and at optimum cost, with ongoing job satisfaction to all those involved in its creation.

New opportunities

The concepts of total quality management provide the professional with the opportunity to bring home to his client that it is he, the client, who must help produce the quality building, first by selecting the quality professional firm with its own quality management system and then by involving himself properly in the decisions which he must continue to make. If collateral warranties are required for any future occupier of the finished building, the risk involved will then be small. Furthermore, by his participation in the project quality management plan and involvement in the project quality system, the client will also be, in effect, a signatory to the warranty.

A client that has utilised an audit to establish the suitability of his design team members should, with their assistance, then establish a quality management system suitable for his specific project.

However, each member of the design team should then be positively involved in the advice and recommendation of prospective construction services. To this end, further study and use of the procedures recommended to carry out audits, firstly for the appointment of a main or management contractor and secondly to influence the selection of sub-contractors and suppliers (with or without design and/or manufacturing facility) will extend the auditing concept to the later team members in a partnership rather than protagonist mode, an essential ingredient of the total quality management philosophy.

The design team should develop a questionnaire similar to appendix 3 to obtain information about any company that it considers employing, in order to ensure that the client's quality, time and cost requirements can be fully understood by the potential tenderers. If the subject company is the right one it will then, through its knowledge, experience and own quality management system, be able to conduct a similar assessment, or auditing procedure, for the selection of its subcontractors.

This continuation of the second-party auditing process will help to establish the need for quality management to be reflected from the client right down to the smallest supplier. It will certainly pay off in the quality of the work, as well as in its cost, with improved human relations and reduced hassle as a bonus.

The success of total quality management in other environments has been encouraged by the joy of work through which everyone wins — a concept of morality projected by several of the gurus but not frequently in the forefront of the minds of contractors and subcontractors.

Recent developments encouraging professional competition through fees rather than performance is perceived by professionals and contractors alike as reducing the standards of professionalism of the design team. The philosophy and principles of total quality management, properly explained and demonstrated to clients, will help to restore professional standards and performance and extend the moral concepts away from conflict towards co-operation and harmony.

7 Total quality management and the contractor pre-contract and post-contract

Where total quality management is used for a project, the contractor should be presented with the framework of the project management plan and manual. He will have his own corporate quality manual and from this he will then develop his own plan and management system for the project. A list of sections (based on contract work procedures developed by a leading general contractor) is provided by way of illustration.

THE NATURE of the construction industry is such that often the contractor will be carrying out some of the design stages, and so many of the implications of a project management system for him are dealt with in the earlier chapters.

The principles relating to project quality are described in earlier chapters. If the client and his agents — project manager, architect, engineer — have followed these, the task of the contractor will be made much easier. He will have been presented with the framework of the project's total management plan and will have been inducted into what is required of him for inclusion in his own part of this plan.

The contractor does, however, still have a major part to play in the overall project quality management system by developing it for the construction phase.

Before he can effectively and easily develop his part of the quality management plan for the project, he will have had to establish his own company's quality management system that will apply to all projects which he undertakes. It should then be a matter of pulling out from the overall company system those matters which must be dealt with by procedures for the particular site and in accordance with the particular rules of contract and to the standards required by the project's documentation thus far. He will therefore probably have established a work procedures manual which may well run to 200 – 300 work procedures. He will have debated and resolved

Table 18 (below, facing page and overleaf). Contract work procedures

---

	Issue and amendment of procedures
	Amendment distribution form
	Form for recording amendments

Section 1	*Preparation for tender*
1.1	Action on tender receipt
1.2	Document received
1.3	Initial discussion check-list
1.4	Letters for declining tender invitation
1.5	Accepting tender invitation
1.6	Notification of tender
1.7	Tender strategy meeting
1.8	Review of tender risk
1.9	Site visit check-list
1.10	Despatch of tender
1.11	Document control and storage

Section 2	*Award of contract*
2.1	Contract award
2.2	Contract procedures
2.3	Authorisation — works order
2.4	Contract documents filing and review procedure
2.5	Contract handover meeting agenda
2.6	Contract planning check-list

Section 3	*General procedures for control of contract documents*
3.1	General procedures
3.2	Tender offers
3.3	Contract documents/agreements etc.
3.4	Incoming drawings
3.5	Consultant's/specialist drawings
3.6	Specifications
3.7	Bills of quantity
3.8	Contract completion
3.9	General notes for filing index
3.10	Site diary
3.11	Archiving procedures
3.12	Filing index

Section 4	*Procurement*
4.1.1	Procement of materials
4.1.2	Vendor lists
4.1.3	Material requisition
4.1.4	Quotation summaries
4.1.5	Despatch of orders
4.1.6	Post-contract orders
4.1.7	Initiating orders
4.1.8	Site orders
4.1.9	Additions to original orders
4.1.10	Bulk orders
4.1.11	Surplus material
4.1.12	Samples
4.2.1	Subcontractor packages
4.2.2	Subcontractor questionnaires
4.2.3	Subcontractor data records

---

continued

Table 18 — continued

continued

Table 18 — continued

Section 5.5.10	*Surplus material	
5.5.11	*Theft	
5.6.1	Purchaser (client) supplied product	
5.6.2	Special material	
5.6.3	*Insurance	
5.7.1	Setting out	
5.7.2	Calibration of equipment	
5.7.3	Hired-equipment requisition	
5.7.4	Records	
5.7.5	Datum and grid lines	
5.8.1	*Inspections and test	
5.8.2	*Testing	
5.8.3	*Test certificates	
5.8.4	*Test facilities	
5.9	Protection and cleanliness	
5.9.1	Protection of finished work	
5.9.2	Cleaning of roads and pathways	
5.9.3	Site cleanliness and clearance of rubbish	
5.9.4	Waste disposal	
5.10	Procedure for handling disputes	
5.10.1	Complaints record sheet	
5.10.2	Investigation of complaints	
5.11.1	Accident report	
5.11.2	Loss and damage report	

* The very limited number of routine inspections of the less important work procedures by quality management personnel rather than by line managers illustrates that effective quality must be built-in if it is to produce a quality building and so a satisfied client.

the argument as to whether a quality manager polices the activities of the site team, reporting separately to a quality director, or whether quality is the line-management responsibility of the site manager, contract manager, site agent, etc.

The guidance given in ISO 9004−2 applying the principles of quality management to a service industry has a long way to go to overcome the misleading emphasis conveyed to the construction industry by ISO 9001−9003, which although ideal for manufacturing situations do not meet the more stringent and specific needs for controlling one-off construction projects on site.

To illustrate this point, Table 18 shows a list of the contract work procedures as developed by a leading general contractor. Indicated with an asterisk are those procedures which a senior contracts director or manager — with limited time available when visiting a project to review its progress with the site manager — would not generally consider a vital exercise of his managerial responsibilities. They may therefore be matters which the quality co-ordinator or quality manager on visiting the site might review to ensure that the procedures are established and working on that particular site. The very limited number of such routine inspections of the less important work procedures by quality management personnel rather than by line management illustrates that effective quality must be built-in if it is to produce a quality building and so a satisfied client.

Total quality management and the contractor's project chain

An area not yet touched on or developed much in practice is that of vendor and subcontractor involvement in total quality management. BP Chemicals recently made a major assessment of its project performance and realised that improvement in future projects would benefit from a benchmark against which to compare vendors, as well as from improved experience in applying certain concepts of total quality management within vendor companies. BP Chemicals are trying to develop more active links with vendors and subcontractors. It hopes that this greater cohesion will produce positive improvement in problem prevention and service development.

Both vendors and subcontractors play a crucial part in any project. The delivery and installation of plant and equipment to time, cost and quality is vital to a project's overall outcome. The ideal would be to have subcontractors and vendors practise total quality management in line with the main contractors. However, it would be inefficient to develop a positive total quality management link with all vendor and subcontractor companies due to the volume of personnel and the limited duration of their involvement. Longer-term goals are not only the concepts of total quality management on site but also should relate to some of the site practices. First of all, site induction courses should develop total quality management exposure for each employee, with selective supervisor training.

Furthermore, vendor and subcontractor assessment must be developed so that its principles are fed back to the company to attempt to create a better service. This should clearly affect a contractor's performance on the preparation of requisitions. Certainly BP Chemicals has already announced that it intends it to be known generally that the criteria for acceptance will rely on meeting not just ISO 9000 standards but also those of total quality management.

BP Chemicals believes that total quality management on projects not only benefits the particular project but also can assist in a continuously improving process. When the company is considering new contracts, subcontractors and vendors will be involved.

From the lessons BP has already learnt, and the new ideas it has been developing and building on its achieved successes, it sees total quality management as a method of improving its overall project performance as well as its corporate performance. It believes this holds a message for industry as a whole, which the author of this book clearly endorses.

The result of this approach, which has also developed in the USA and Australia in the last 2—3 years, is known as Partnering, and will lead logically and sensibly to a considerable reduction of the number of vendors and subcontractors that are used by a general contractor, where it has the choice. This cannot always be so, particularly in specialist contracting niches where the client's specific requirement may be only capable of supply by one or two firms outside a general contractor's normal 'domestic' range of subcontractors.

The likelihood of a contractor forming partnerships with traditional craft-based subcontractors and suppliers is probably greater. However, success would be reliant on the volume and continuity of work, a particular problem with the stop—go use of the industry as an economic regulator.

The opportunity is there for subcontractors and vendors to take the initiative and develop total quality management themselves rather than relying on clients to provide the means. This would be an important and impressive step. Total quality management is the framework for effective project teamwork by all parties. This emphasises again the differences between quality assurance (and ISO 9001 and ISO 9002) and total quality management.

8 Project quality management for the specialist subcontractor

The quality required by the customer in construction is in the end product, the finished building, not in the separate quality management systems of the participants. Manufactured products account for only 10% of building failures. Thus, for specialist subcontractors who also install their products, product manufacturing systems are only part of the story. Nevertheless, ISO 9001 has its part to play. This is outlined, but set in perspective against the needs of total quality management on projects and the importance of the project contract and specification. Third-party certification and the general requirements of ISO 9001 are included.

THE DEGREE to which project procedures and total quality management will affect a specialist subcontractor will depend partly on the degree to which the subcontractor manufactures standard products which he then installs on site against specific contractual requirements.

Where the subcontractor's process is one of repetition of manufacture of a product the design of which is also his responsibility, it can be assumed that his production line will have been the subject of a quality assurance system developed in accordance with the guidance given in ISO 9000, and ISO 9004 − 1, to comply with ISO 9001 (BS 5750 part 1). Certainly, if the product is manufactured in the UK he is likely to have had it third-party certified by an agency accredited by the National Accreditation Council for Certification Bodies, (NACCB) thus assuring purchasers that the product itself comes from a production line that assures uniformity and therefore that any kitemark or other product conformity certification can be assumed to apply to all products delivered to site. The way in which the ISO 9000 series applies to the different sectors of the industry, and also the difference that ISO 9004 − 2 provides for service industries is included in Table 16. The general implications of third party auditing in relation to the requirements of ISO 9001 − 9003 which were developed for and by manufacturing industry is now very relevant for the subcontractor who manufactures and installs or erects.

The purpose of ISO 9001 – 9003 is to bring reliability and economy in the production process. The system controls the process from start to finish and provides a complete record of every stage of production. This is invaluable for product or process improvement and in relation to any product liability claim.

Suppliers can use ISO 9000 when setting up their own quality systems. Customers may specify that the quality of goods and services they are purchasing shall be controlled by a management system complying with ISO 9000. Customers or third parties may use the standard as a basis for assessing a supplier's quality management system and thus his ability to produce satisfactory goods and services.

The objective of having the system assessed by a third-party agency is to enable purchasers of the product to know, without themselves inspecting the supplier's system, that the system exists and is used to assure satisfaction of the product against both product specification (if there is one) and system specification as prescribed in the ISO standards. If the certifying agency has been itself accredited by the National Accreditation Council for Certification Bodies, then this ensures that the agency has its own credibility through systems, procedures and people competent to deal with the area of industry set out in the scope of their accreditation. (An accredited agency is not accredited outside of the scope of its accreditation, although some do certify systems outside their particular area of industrial expertise.) Third-party certification against ISO standards is, as yet, much less practised outside the UK than within it.

ISO 9001

Section 4 of ISO 9001 has 20 principal headings for the quality assurance system. This pattern is followed with appropriate amendments in ISO 9002, where design is omitted, and in ISO 9003, which is limited to final inspection and test. Summaries are given below but should not be used except as examples. The standard should be referred to for working purposes.

Section 4.1. Management responsibility. Important prerequisites to commercial success are company policy and objectives. These must include quality, and like all systems ISO 9000 requires good organisation to implement them. Responsibility must be identified and assigned to all personnel who manage and perform work affecting quality, and trained personnel must be assigned to verify that the work is performed properly. One manager with the necessary authority and ability must be clearly put in charge to see that prompt and effective action is taken to ensure that all the requirements of ISO 9000 are met.

Here lies one of the first differences between manufacturing, where machines do the work, and services such as site installation, where people do the work, and so where all are concerned with quality and all must have their responsibilities defined. Review of the quality system is essential and must include internal quality audits. The review should reveal defects or irregularities and check on the effectiveness of management at all levels to ensure that management objectives and methods are achieving the desired results. The benefits of these reviews have a considerable bearing on profitability and customer satisfaction.

Section 4.2. Quality system. The nature and degree of organisation, structure, resources, responsibilities, procedures and processes are essential matters for the system. It is important that they are documented so that they are readily understood by appropriate personnel and the quality system is maintained at a level to provide constant control of quality.

The quality system must be planned and developed to take into account all other functions. Quality planning must also identify the need for updating of quality control techniques, ensure that equipment and personnel are capable of carrying out the plans, and provide for adequate quality records.

Section 4.3. Contract review. Before commencing work, it is essential to review the contract to ensure that the contract requirements are complete and not ambiguous, that they do not differ from the original enquiry or tender, and that the resources are available to meet the requirements. Contract review in construction may need to be co-ordinated with the customer's organisation.

Section 4.4. Design control. It will be necessary to establish and control the functions of design, assigning activities to qualified staff with adequate resources, controlling interfaces between different disciplines, documenting design input requirements and documenting design output and calculations. Competent personnel will need to be assigned to verify that the design output meets the design input requirements and to control all design changes and modifications.

Section 4.5. Document control. A co-ordinated system will be needed that ensures that all the appropriate documents are available wherever they are essential and that changes are authorised by the originator of the document. Also written procedures are needed which describe how functions shall be controlled, what is to be controlled, where and when, and by whom.

Section 4.6. Purchasing. Nothing is more aggravating than the failure of bought-in product. Control should be exercised — in writing — of purchased product and services, and purchasing data. Thus purchased product should be inspected and verified, together with the quality system to be applied, as appropriate, by the suppliers.

Section 4.7. Purchaser-supplied product. The suitability of purchaser-supplied product for its purpose must be assured (not assumed). In addition, the specialist subcontractor is responsible for freedom from defect, maintenance, storage, handling and use while the product is in his possession.

Section 4.8. Product identification and traceability. Product identification on an item basis provides important protection where products differ but the difference is not visually obvious. Bought-in product released for manufacture prior to evidence of conformance being available is another key area where identification is essential. Where traceability is specified for safety, statutory or other reasons, documented product identification provides the capability to trace and recall non-conforming or hazardous product delivered to a customer.

Section 4.9. Process control. All manufacturing operations — including written work instructions — must be carried out under controlled conditions. If any operation or process is excluded or missed, the result may be below-standard products. Good work instructions cut out confusion, and show what work must be done or services provided. Procedures and work instructions should cover every phase of manufacture, assembly and installation.

Section 4.10. Inspection and testing. Procedures for incoming inspection need to take account of the documented evidence of conformance provided with incoming goods. All procedures should consider techniques, the accuracy and suitability of test equipment, and the type, competence and accuracy of the data to be recorded throughout the process, and all non-conforming products must be identified.

Section 4.11. Inspection, measuring and test equipment. Suppliers must provide, control, calibrate and maintain inspection, measuring and test equipment to suitably demonstrate that their products conform to specified requirements. It is important that the calibration status of equipment is traceable to nationally recognised measurement standards and that equipment has the necessary accuracy and precision consistent with the capability needed to measure the product. Procedures and records covering the control of the equipment are required.

Section 4.12. Inspection and test status. Suppliers should establish a system for identifying the inspection status of product during all stages of manufacture. Written control procedures are necessary so that it can be established quickly at all times whether product has

- not been inspected
- been inspected and approved
- been inspected and rejected.

The procedures should provide for records of the inspection authority responsible for releasing conforming product for further operations, assembly or dispatch.

Section 4.13. Control of non-conforming product. All non-conforming product should be clearly identified to prevent unauthorised use, shipment or mixing with conforming product. The documentation should record the identity of the product, the nature and extent of the non-conformance, the authority for the review and the decision on disposition (including how it should be carried out). If it is to be reworked, reinspection should follow documented procedures.

Section 4.14. Corrective action. Effective corrective action is essential. Segregating defective product is not enough: the cause must be identified. Incorrect working methods are frequent causes of defects. When faults are discovered, the unsatisfactory design specification and work methods should be changed.

Section 4.15. Handling, storage, packaging and delivery. Written instructions and procedures should govern the way product is handled and protected in process and should ensure that special product is not mixed with similar product of unknown or dissimilar quality, that there is no contamination, that delicate parts are protected and that product does not miss any operation or inspection.

Section 4.16. Quality records. Records are the objective evidence that quality requirements are being met. They should encompass audit reports of the quality assurance system, results of inspections and tests, calibration of test and measuring equipment, approvals, concessions and corrective actions.

Section 4.17. Internal quality. Overall functioning of the system is monitored by management through the assessments of the results of internal quality audits. These audits should be in accordance with a documented procedure to verify whether quality activities are performed as planned.

Section 4.18. Training. Activities demanding acquired skill should be identified and the necessary training provided. Competence should be demonstrated by examination, testing or certification. Records of training and achievement must be kept.

Section 4.19. Servicing. Where servicing is normally provided or required by contract, suppliers must establish procedures for controlling and verifying the quality of services performed.

Section 4.20. Statistical techniques. Statistical procedures should include the manner of establishing process capability, identifying lots and clarifying characteristics.

Conclusion

The introduction to each standard emphasises that the quality system requirements specified in ISO 9001 − 9003 are complementary and not alternative to any technical product specification, and refers to ISO 9000 for guidance on tailoring the requirements for particular contractual situations.

Some manufacturers unfamiliar with the ways of construction have found that their systems for construction projects and approach to such projects are hindered rather than helped by trying to adapt the language of the manufacturing quality systems to the site. As stated in chapter 3, for repetitive processes the ISO standard should be the basis of the system and its auditing and assessment. For unique projects, total quality management and the project's specific requirements and standards should be the basis of assessment and auditing.

Thus the best references in these circumstances are the preceding chapters rather than the ISO standard. Remember the bold reminder on the inside of the front cover of the standard

> 'Compliance with the standard does not of itself confer immunity from legal (i.e. contractual) obligations.'

9 Total quality management during commissioning and maintenance

This chapter deals with a most important and often difficult stage at the end of the construction process. With the complexity of modern buildings, the time and cost of commissioning are too frequently underestimated or, indeed, ignored. Only now are engineers establishing a specific role as 'commissioning engineer'. This role is made more necessary by the dispersed nature of services design between consulting engineers and specialist contractors.

THE LACK of completely integrated working drawings for the entire plant before its installation begins causes strains which fall first on the construction management in its various forms. Construction management is extremely hard pressed to field M&E managers or contract engineers with sufficient technical skill and experience to cope with these demands. Experienced construction management (and experienced consulting engineers and clients sometimes) are now seeking the assistance of a specialist commissioning engineer to unravel the often dire and unsolved complications running through the installation phase into the commissioning work that imperil completions and handovers.

Commissioning engineer

The capacity of the supervisory commissioning engineer to catch the problems early enough in the project programme and, without contractual authority, to persuade the construction management, the consulting engineer and often the client to change course is crucial. It usually requires a whole-man-engineer of considerable breadth, skill and experience, undaunted by specialists and by demarcations between engineering disciplines and categories.

Unfortunately the role of this commissioning engineer (who may well be the only person on the entire project who understands in every detail the way in which the plant is either supposed or likely to work) has been widely misunderstood, distorted and undermined. Instead of being carefully recruited in a highly selective manner as a professional engineer, more often than not he is bundled into a subcontract or trade contract package and the contractor is invited to submit a lump sum tender in competition with others for work which by its very nature is impossible to quantify at the time of tender.

Total quality management on a project basis must develop earlier attention to this integration if the commissioning phase is not to be a long series of rejects that can only lead to long disputes.

The difficulties are particularly acute at the interfaces between the general mechanical services contractor and the specialist manufacturers for air-handling units, refrigeration, boilers, generators, building management services and life safety, and so on. This is an industry-wide problem and is not confined to projects in which the specialist engineering is in separate work packages from the general engineering. If the Engineer has delegated application design to the specialist, the problems are just as acute as if the specialists are subcontractors of the M&E contractor. Where the work is in the hands of an overall M&E design contractor, the situation is likely to be even worse due to his lack of design expertise in these specialist fields. Where a work package contains no delegated application design responsibility there is less difficulty even at the interfaces.

Tuning the plant

When all the individual parts of the M&E system have been correctly installed, commissioned and calibrated to their design settings, the construction management, the contractors and sometimes the client are inclined to equate the situation with 100% success and completion. Such optimism is rarely justified. Most installations need tuning to optimise all the adjustments on a great number of technically interactive but commercially independent packages of equipment. This is more work for the whole-man commissioning engineer, assisted by specialist contractors' staff when necessary. However, there is very rarely any provision at all either in the project programme or in the cost plan for any such work. This essential last link in the project engineering chain before handover to the client's operation and maintenance crew is often missing altogether.

However, if the total quality management procedures have been followed since the commencement of the project and the client has been involved throughout to the extent necessary, the handing over and procedures connected with it will be more of a routine and paperwork formality.

At each stage each service should have been signed off by its provider after testing, and its status documented. Good management and teamwork developed by good leadership should have helped to ensure that damage to unconnected services has been prevented. There will remain, however, a whole series of services that can only be connected, debugged and balanced when everything has been completed in the construction phases.

On a ship, such activities are done over some weeks or months: contractors, engineers and owners' officers sail together to make sure that all is in working order and performs according to specification before the commissioning is complete and the handover is signed in acceptance. A similar period of proving flights applies in the case of aircraft. Too often a building owner expects to occupy and find everything satisfactory on day one. Alternatively, he spends the whole of the defects liability period under the contract finding out the problems, then, years later, has them rectified. The final payments calculated by the quantity surveyor are then withheld pending litigation.

Total quality management of the project, commenced at its inception, should ensure effective co-ordination of all the application design work, and see that it is concluded before the start of any installation work. Thus most problems would be resolved or prevented, but proper provision for commissioning would still be necessary.

The services contractors will find it much easier to keep pace with the overall construction programme if the key factors are observed (and monitored and documented). This includes

- complete resolution of the design interfaces of the work packages
- completion of the entire-application design drawings in working form prior to the start of installation work

- single point of reference for all queries on plant technology
- whole-man engineer in collaborative support of the construction management and the contractors engaged in fabrication, assembly and commissioning.

It may be that this will take time, not least to ensure that the capacity exists for engineers to fulfil this role. But a strong collaborative client who understands the need and demands it (rather than the sum of many lowest tenders from consultants and contractors) should see that the response will be more cost-effective and less confrontational and litigious.

During the commissioning for each of, say, 15 services, the following steps will need to be gone through

- subcontractor's testing, documentation of test mandates, and handing over to the general contractor
- general contractor's confirmation and documentation of the handover
- engineering consultant's confirmation and documentation
- Architect's confirmation and documentation
- acceptance by the client's facilities manager.

After all the services have been commissioned and accepted (and perhaps after all other snagging lists have been cleared off), the final inspection of the building takes place. Then the certificate of practical completion can be issued, and the release of retention monies (under the Joint Contracts Tribunal forms of contract) can take place.

Included in the handover for each service there should be a list of all service and maintenance instructions, with copies of them, unless the service — for example, of the lifts or escalators — is to be maintained by the installation contractor. Even then the client will need operating instructions and guidance on maintenance and servicing schedules. Copies of these documents will be needed up and down the line and so this final stage of project documentation can be seen to be approaching 500 sets of documents — assuming there are no repeats after rejection and no phased handover of different parts of the project!

Is any further case needed for an effective project management system? Clearly no third-party assessor making assessments against the requirements of ISO 9001 — 9003 in the case of individual firms who are involved in the project can do anything to organise project quality at this (or any other) final stage. Quality projects must come from total project quality management, not from ISO 9000!

10 First-party auditing

The assessments that an organisation makes of its own activities carried out through a management system is the most relevant and significant act in maintaining and improving the performance of the organisation. The first audit, to establish where the company is in terms of procedures and their documentation in a co-ordinated management system, is known as a State-of-the-art audit. This is a worthwhole exercise prior to commencement of any quality management development programme. It should aim to establish the parameters for the system, and the interfaces between departments and processes that should be controlled. A diagram should be established of the overall processes carried out by the company. At the end of the programme, and at least annually thereafter, review audits should be carried out — always by someone not responsible for the activity being audited. The kind of paperwork involved in this process is illustrated.

THIS BOOK may well be read by someone whose organisation is and has been an official provider of buildings which have satisfied their client in every respect. In a small organisation this may not have been the result of any documentary or formal quality system. In a larger company, customer satisfaction may have been provided through procedures which are well known and well tried, but here too they may not have been documented; or, if they were documented, they may not have been incorporated into a total management system. Book-keeping, accounts and the financial side of the company may well have been completely separated from the contract management, which in turn may have been completely separate from the marketing, at least in terms of documents within a single management system.

Every organisation, whatever its size or function, however, does have four fundamental areas of activity which should be co-ordinated into a single overall system which will facilitate management control to produce more effective performance as a result of benchmarking and auditing the total system. These four areas are as follows

- marketing — everything to do with getting the work
- production — everything to do with carrying out the work
- administration — everything else to keep the organisation fully and effectively operational
- strategy — the principal concern of the directors of the company.

The project management process which requires the management systems can be identified and its flow can be charted as a map of both the paper-activities and the people-activities. Figs 17 and 18 illustrate the interconnection and interfaces between the internal functions and their external interface with the customer and supplier. Fig. 18 also shows how it is possible to identify paperwork systems and procedures of the main process flow with various documentary record books. It shows how it is possible to key the operational procedure with documentary records and at the same time to separate paperwork flows from main process flows. Such a diagram would be an invaluable aid in a quality manual to show new members of staff or auditors strange to the system how it works and where to look for the records to be audited.

State-of-the-art audit

With a diagram such as Fig. 18, developing a total quality management system which will embrace the techniques of quality will be relatively simple. A first audit of the management system can start off a programme to develop a total quality management system. This is a 'State-of-the-art' audit.

In few organisations without a fully documented system will any one person know all of the procedures or forms and charts that exist. Therefore, the first task is to collect (root out might be a better description) all documents that exist and label them with their source and purpose. It will also be helpful to get the person who uses the form or document to note when and how it is used, what initiates its use, where it goes afterwards, who gets copies, where they are filed, and so on. If a trail can be established, the form can be marked off on the process trail: the system (by way of several subsystems) is well on the way to being established.

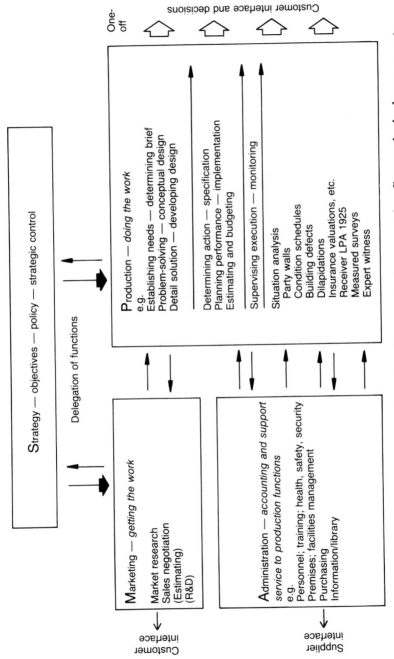

Fig. 17. Process flow diagram through a functionally organised construction firm: principal components

Fig. 18 (below and facing page). Identification of activities and documentary recording of procedures through the overall process

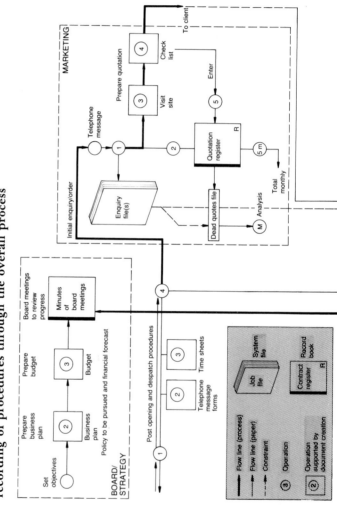

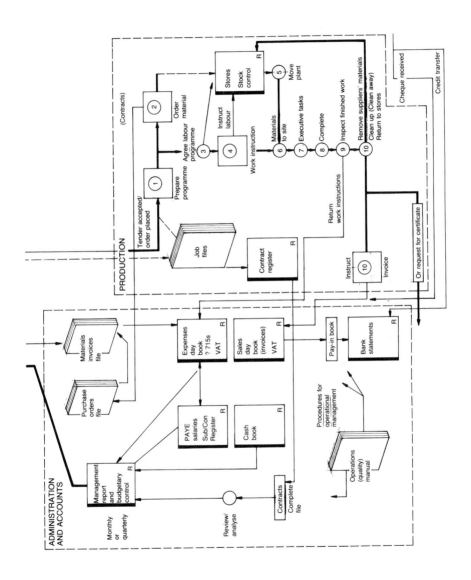

In principle, all that is needed then is to

- establish the gaps in the system
- identify the linkages between subsystems
- create procedures and/or forms to bridge the gaps.

However, in the process of this voyage of discovery many surprises, both as to what does exist and what does not, will emerge. These will in themselves identify improvements that can be made in the system by elimination of unnecessary overlaps, and productivity is already under way.

In the course of this voyage of discovery, all those involved should be encouraged to join ship and make their own suggestions. Sometimes the suggestions will improve their own part of the operation at the expense of someone else. It should be remembered that the aim of the system is to optimise the overall performance and not provide sub-optimisation in individual areas at the expense of others.

In the course of this activity to establish the state of the art, it is a good idea to ask several people who carry out any undocumented procedure to write down what the procedure is and how it is carried out. If three or four people are asked to do this exercise independently, there will be at least two or three variations in their description. One of these will certainly be better than the others, and once again an improvement in the system and overall performance can emerge.

However, for the purpose of the initial state-of-the-art audit it is enough to list the forms and chart the procedures in order to develop an overall programme. The development of the system may take anything from three weeks to three years, but the start is with this particular first-party audit.

Review auditing

First-party auditing will, however, continue regularly thereafter. When the system development programme has been completed and is up and running, another first-party audit should be carried out and the system documented and reported to top management.

The first principle involved in first-party auditing is that the audit should be carried out by someone, or a team, that is not responsible for or directly involved in the system, or part of the system, being audited. The role of the audit team is to monitor the conduct of the system and this can really only be done properly by a 'third party' to the system's user.

In small organisations where 'matrix' management is involved, this may not be completely possible. However, in these circumstances the more formal principles are of less consequence than the regular review of the procedures, with the principal purpose of seeking ways to improve them — or, more particularly, the performance of the organisation through their use.

Many firms in the construction industry will have several offices or locations, and a sensible procedure would be to get the people using the system in one office to carry out the first-party audit in another location.

On the first occasion, after the establishment of the system, the auditing may take anything up to $2-3$ times the time that subsequent audits are expected to take. If the main aim is improved performance, the time involved is of less significance than the benefits obtained. Certainly, first-party audit should be more concerned with improvement of the system than with meticulous, pedantic or slavish subservience by its users.

Over-documentation

There will generally be a tendency when establishing the first overall quality management system to produce too many forms and document too many procedures in too much detail. This should be resisted. Perhaps when the second routine audit is carried out, one of the objectives should be to see how much of the system can be streamlined without affecting performance.

For those who wish to, or have to, produce a third-party certificate from an accredited agency to satisfy their market's requirements, and have carried out their own first-party audit, there will be no difficulty in obtaining the certificate.

The purpose of an individual firm's system is to enable it to carry out its part in a project and therefore fit within a project total quality management system, rather than to relate its internal procedures to the requirements of ISO $9000-9003$, which were created and are directly appropriate for manufacturing and multi-repetitive services.

TQM 1st PARTY AUDIT AGENDA
(C.042 - WD 1.2)

Date of Audit: 28/ 05 /92

Start time: 05.00

Sequence No. 1 (M)

Location: *A G Manchester Operations* Person in Charge: *J Kelsall / K Hughes*

Nature and Scope of Audit: 1 *To verify the completion of the TQM System and its*
practical experiment in all areas of the practice in Manchester
2 Identify areas for improvement in the system and
note any areas of non-conformance

Points to be taken with Person in Charge *General degree of satisfaction at present ?*

AUDIT AGENDA

A. MANUALS - to be checked. Nos. *Master - in computer and all section leaders*
(NUMBER ISSUED -)

Up to date/amendment needed

Corrective Action needed
at last Audit on   /   /

	C/MR	D/1						
i)		C·601						
ii)		C·602						
iii)		C·301						

Time

B. **Opening Meeting**

Persons to be seen:

✳ B Ingram	9.45
D Sleigh	10.15
✳ J Bradley	10.45
J Cundliss	11.30
✳ K Hughes	11.15
✳ R Grogan	12.00
P Harvey	2.00
R Smith	2.30
✳ W McNeil	3.15
✳	
Closing Meeting	4.30

C. **Management Responsibilities/Mandates**

Is the structure up to date?

New ones required

Amendments

Management Meetings

Nomenclature not consistent e.g
L/OM, P/A/S in charge
Cost centres - cc manager's etc
Leaves doubt on responsibility sometimes
from lack of complete system, sometimes
from lack of clarity in wording

Fig. 19 (a) (above and facing page). First-party audit: agenda

D. PROCEDURES to be checked in detail:		NOTES/CORRECTIVE ACTION NEEDED
1. General Procedures	C/AN	
i)		
ii)		
iii)		
iv)		
v)		
2. O/A administrative Procedures	C/AN	
i)		
ii)		
iii)		
iv)		
v)		
3. Departmental Procedures	C/AN	
i)		
ii)		
iii)		
iv)		
v)		
4. Personnel Procedures	C/AN	
i)		
ii)		
iii)		
iv)		
v)		
5. Other Procedures ?	C/AN	
i)		
ii)		
iii)		
iv)		
v)		
6. Project Procedures	C/AN	
i)		
ii)		
iii)		
iv)		
v)		

Other members of the Audit Team: 1. ....   2. ....

Date Completed   /   /   Signed: ....Copies to: ....

TQM 1st PARTY AUDIT REPORT FORM   Date of Audit: 28/ 05 /92
(C.042 - WD 1.3)

Start time: 09.00 – 17.20

Sequence No. I (M)

Location: T.G. Manchester Operations   Person in Charge: J. Kelsall / K. Hughes

Nature and Scope of Audit: 1. To verify the completion of the TQM System and its practical application in all areas of the practice in Manchester.
2. Identify areas for improvement in the system and any areas of non-conformance.

AUDIT TEAM (Leader)   R B H   2. J. Kelsall   3. Kevin Hatton

A. Points taken with Person in Charge   Range of satisfaction for 20% - to 60%

MANUALS - checked.	Nos.	Master – in computer and all section Leaders
		(None numbered)

Corrective Action needed		C/MR								
(Non conformance)	i)	B.2.09	C.601							
	ii)		C.602							
	iii)		C 301							
	iv)		C.304							
	v)									

B. Persons seen: **Time**

* B. Ingram   9.45

D. Gwyn   10.15

* J. Bradley   10.45

J. Gardiner   11.30

* K. Hughes   11.15

* R. Grogan   12.00

P. Harvey   2.00

R. Smith   2.30

L. McNeil   3.15

* Attended
closing meeting   4.30 to
5.15

Comments - general ① The manual is said to be "mandatory" throughout. But some things are only for guidance. Some wording is inappropriate for the levels at which some of the procedures will be used.

② The master copy in the computer is not yet in manual classification order, so it is not possible to check it properly.

B.2.09 - Secretary job specification not in the hard copies inspected.

③ Manuals have been kept "virgin" they should be seen as a working tool.

C. Management Responsibilities/Mandates

New ones required

Amendments

Nomenclature not consistent e.g
L/OM, P/A/S in charge
Cost centres, cc managers

Leaves doubt on responsibility sometimes from lack of **complete** system, sometimes from lack of clarity in wording

Secretary job specification held in computer not distributed in hard copies

Fig. 19 (b) (above and facing page). First-party audit: report form

D. PROCEDURES to be checked in detail:	C/AN	NOTES/CORRECTIVE ACTION NEEDED - ACTION DESIRED	A/D
1. Management Meetings		agendas should now have "Quality System" headings in Production and Management meetings	
2. General Procedures C/6.02, C/6.03, C/6.01 C/3.01, C/3.04	C/AN ✓ ✓ ✓ ✓	Signing out - not complied with Telephone recording - poor discipline complained of Owner's time sheets (MANDATORY) not completed Monthly budget update not done	
3. Administrative Procedures None written	C/AN	Not yet developed "Contents list" urgent Until this exists and procedures are written whole system can not be verified	✓
4. Departmental Procedures C.4 Mechanical Services	C/AN	Not yet developed	✓
5. Staff Procedures H.A.S.W.A Grievance + Discipline	C/AN	These are difficult to find due to mis-change can under "Owner's Procedures"	✓
6. Project Procedures	C/AN ✓	Regrouping under more appropriate classification essential - see Polycom's wider comments	
7. Other Procedures ? Owners Procedures C.3	C/AN	"Owner's Procedures" is not a sensible main heading they are either "Responsibilities" or should be dealt with under appropriate Project or Admin procedures	
General Recommendations/Comment Overall Classification • Mechanical Services • Project Management • Site Workers • Special Sector Variants • Admin + Accounts	✓ / ✓	Tutoring - Regrouping changing some classification needed to make use of manuals easier Should all be identified and titled more likely and procedures then drafted by appropriate Section Leader for verification and discussion with Quality Partner	

Other members of the Audit Team were:- 1. John McGrail   2. Kevin Hutton

Date Completed   /   /   Signed: ....   Copies to: J.K. S.H.

The kind of paperwork that a first-party audit will generate is shown in Fig. 19.

The British Standard guide to quality systems auditing is BS 7229. It was prepared under the direction of the Quality Management and Statistics Standards Policy Committee in conjunction with the programme of work of ISO Technical Committee TC/176 — Quality Management and Quality Assurance (see Appendix 2).

This standard provides further guidance on the auditing of quality systems, and sets out audit procedures, criteria and requirements for good audit practice and systems. It deals with the planning and execution of audits of quality systems. It has been prepared in a general way so as to be applicable to different industries and organisations, and it defines in detail audit objectives and responsibilities, audit execution, the preparation of an audit report, and how corrective action and follow-up should be taken.

However, the author's view is that, like ISO 9001 − 9003, it is an excellent document when the audit is specifically concerned with the manufacturing and multi-repetitive service industries, but it is not well related to the needs of the construction industry and its customers.

First-party or second-party auditing

The benefits of auditing in the construction industry will be best achieved by the process of second-party auditing, or auditing on behalf of a second party, which is specifically project-related.

Construction projects frequently involve capital investment equal to many years' output from a manufacturing organisation. It is therefore important to apply the auditing principal to the parties who will be involved in a project but to limit the audit to the extent of their involvement in the specific project and to carry this out before a contract is entered into.

The project contract, with whatever type or size of organisation, will be to fulfil a one-off activity rather than to continue to supply hundreds or thousands of products over a long period of production. These audits are dealt with in the next chapter.

Self-assessment

Chapter 1 identifies and outlines the framework of the Baldridge Award and the European Foundation for Quality Management Award, and forecasts a new British award from the British Quality Foundation. All of these approaches lay great stress on self-assessment as an important and essential element of the management of quality. Any management which leaves the review of its performance to an outside agency on an occasional basis is clearly not practising total quality management within its own organisation and is clearly not likely to be ready to contribute to total quality performance on a client project within the construction industry.

The principles of total quality management delegate responsibility to the lowest possible level following the X and Y approach of McGregor. In contrast, a maxim of quality assurance is that somebody other than the person carrying out the task should check that it has been properly done. Where calculations are concerned in a professional activity, the balance between verification by another and responsibility by the doer is critical. As a generalisation, the greater the motivation, the less the need for monitoring, but in really critical areas the prevention of human error will take precedence.

Within an operating team the 'big brother' type of policeman has a minor role, but in total quality management first-party assessment is vital. First party assessment will always be far more effective than the introduction of an external third-party policeman, which is a denial of management authority, leadership and good practice.

11 Second-party auditing pre-contract

Second-party auditing has a dual role: it illustrates for all those involved in the project that total quality management has been developed at an early stage in the client organisation, and will be maintained throughout the project; and it seeks to establish an affinity with it from those firms who will be involved in the project's execution. This chapter sets out the purpose of and approach to pre-contract audits. Appendix 3 gives a model questionnaire for use by a client or his design team.

A S COVERED in chapter 10, all those involved in the building process from client to sub-subcontractor should be maintaining their own organisation's quality management system by regular first-party audits. However, from the client's point of view, it is the project that must have an effective quality management system and this must be developed early. Project auditing should be undertaken by the client, or on his behalf, prior to the appointment of, or signing of a contract with, any of the firms who will make up the project team (Fig. 20).

Second-party project-specific auditing must illustrate the essential maxim for quality management — that the philosophy, commitment, attitudes, practices and procedures must start at the top, and only then will they permeate down through all levels of the organisation. In the construction project the top is, and must be, the client for the project — he who will become the second party to all contracts entered into: first with the design team, next with a main contractor or contractors, and sometimes with subcontractors for specialist work.

The client should appoint a 'client representative' and then a project manager, either from within his own organisation or as a further contractual relationship for which a second-party audit is desirable — if not even more essential. But it is the client who must determine the key factors for the project: what the project is, where and when it is to be built, to what

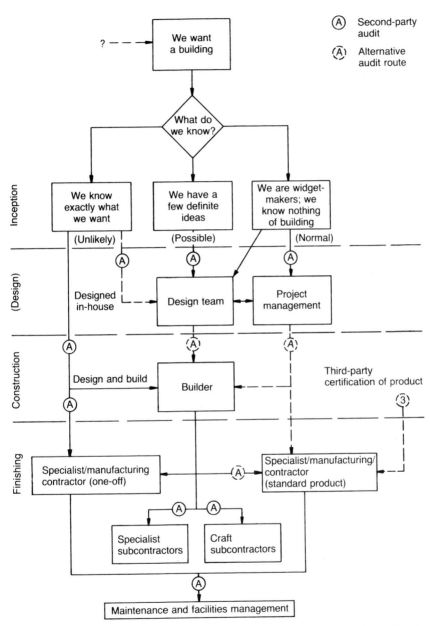

Fig. 20. Location of pre-contract (second-party) project-specific audits in the client alternative route through the building process

standard and order of cost it should be built, and finally who will be selected to carry out the work. He may delegate much, or even all, of the work but the final decisions must remain his.

The prime purpose of these auditing procedures is to see that the organisation selected to carry out the project, or their part of it, is itself equipped to undertake the task and can provide a reasonable assurance that its resources and management will meet the project's requirements and client-defined standards in every way. This is not just a matter of documentary procedures but also of people-systems and perhaps the organisation's motivation, which should then be reflected in the contract that it signs.

The questionnaires to be used seek to establish whether the procedures and systems exist, and to what extent.

Preparing the preliminary information
Inception stage
The efficiency of any project or organisation starts with top management and in the building industry this is effectively the client. Building clients are often 'amateurs' in respect of the construction of their project and rely heavily on their appointed consultants to steer their project to a satisfactory conclusion. The traditional organisation of the building industry has always had regard to this inexperience of the client. Indeed, the pattern of the industry and forms of contract which have grown up have this at least as a background *raison d'être*. The effective overall management of a building project is therefore a task for someone other than the client, and will involve knowledge of all the management techniques needed to achieve the client's objectives, and a quality building. It follows also that the appointment of the 'management team' is one of the most important decisions to be made by the client in the initial stage of the project.

In the past, the client has expected these skills from his architects, or engineers, but has also sometimes shared out these responsibilities between other professionals involved; thereby flying in the face of the principles of delegation of responsibility in single line management. Increasingly, these days, with complex projects, management companies have been appointed to 'project manage' on the client's behalf.

Whichever procedure is adopted the first task will be to establish the terms of reference which will be effectively the brief for the project.

This brief must first be obtained from the client and then examined to ensure that it presents a feasible exercise. Only when this has been established can the design process begin.

Client instructions

Instructions to prepare designs for a building project may arise from a wide variety of circumstances. They may be well- or ill-defined in the client's mind. Either way, it is a task of some magnitude to extract them from a client and understand what he wants the project to achieve.

More often than not the requirements, as first presented, will conflict with one another, and particularly so where a number of people in the client's organisation have individual requirements. In the design of a hospital, for example, there are a multiplicity of clients, consisting of the management board plus each medical consultant and surgeon, and the administrative and paramedical disciplines, never forgetting Matron and the various trade unions!

If the client has established his objectives for the project in a systematic way, he will be in a far better position to pass the information on to those who will need it. However, the process involved of analysis by question and answer must be recognised as the most important part of the exercise, and one to which adequate attention must be given before information transfer is attempted.

The degree of relative importance of the four factors — technical, aesthetic, cost and time — will vary with each project, and in some projects any single one of them may be completely unimportant relative to the others. Often, when the client's requirements have been stated it will be seen that there is clear conflict between functional requirements and cost. It is therefore the first task, once the initial brief has been taken, to assess the project's feasibility. If the instructions appear to so conflict as to jeopardise the project on grounds of either cost or time, it is the consultant's duty to so report to the client and get the requirements amended until they do become feasible. As a general rule, it can be said that at least one of the four factors must be open-ended.

Feasibility study

The first brief should generally not attempt to contain more information than is necessary to report back on the feasibility stage of the project. The client and his design team can then proceed in the knowledge that the project is real and attainable, requires redefinition or should be aborted before costs escalate.

The brief document

The brief document is perhaps the most important document prepared in connection with a building project. As a result of study of the requirements of the brief, when completed, the client will make the decision to proceed with what is probably

Table 19 (below and facing page).   Scope of project for second-party audit

---

Outline
1.1. Outline intentions
1.2. Relationship of the project to other developments or existing buildings, including stages if relevant
1.3. Gross areas of development, including any circulation space and external features
1.4. Target dates for commencement and completion, including any punitive penalties likely
1.5. Gross costs, if necessary broken down into building costs, fees, removals etc. (transmitted only to certain parties selected by the client)
1.6. Any outline or detailed planning permission details

Location
This section should be as precise as possible. It will certainly have a considerable effect on the contractor's and subcontractors' response.

2.1. Site location plan indicating other developments if known and any relevant physical features
2.2. Does the site have existing building for modification/demolition? Are ground conditions known? Has any soil survey been undertaken?
2.3. Access on to site, including any restrictions. Are any road closures envisaged? Is parking in the vicinity of the site restricted/permitted?
2.4. Are adjacent property/site owners aware of the proposed development? Are party wall structures involved? Will noise/dust be a problem?
2.5. Are there any height restrictions for plant/temporary building etc.?
2.6. Are there any existing site services (gas, water, electricity, soil, surface water, telephone)? Are any of the existing services overhead? Is their location precisely known?
2.7. Are there any public footpaths/rights of way across the site?
2.8. Has any form of linear survey been undertaken? Is there an OS benchmark adjacent to the site? Are any trees or other features to be protected?

---

the largest single expenditure in his lifetime or in the company's history.

It is the loose way in which this brief-getting process is often tackled which has frequently led to many dissatisfied clients and frustrated architects. When variations have to be instructed as a result of inadequate instruction at the brief and basic design stages, one finds the general pattern of muddle and inefficiency which has dogged the site efficiency of the whole building industry.

Organisation structure

The establishment of the right organisational structure — to ensure that the right people are grouped together in an

Table 19 — continued

Aesthetic
3.1. Are there any particular requirements on the appearance of the building?
3.2. Are there any specific requirements in respect of the quality standards required?
3.3. There will be a quality management plan that is to be followed throughout the project

Cost
4.1. Questions and statements in respect of cost are normally extremely sensitive and it is generally accepted that the design team should be given the basic cost levels within which they should work; the onward transmission of cost parameters to the building team will largely depend upon the type of contract being envisaged; however, any penalties because of loss of trading or manufacture should be identified to all parties

Time
5.1. The timing for the start and finish of a project can be critical; a programme showing desirable/critical dates should be made available; this should include any restraints, if known, and commissioning time if important
5.2. Apart from penalties for non-completion, are there any bonuses for early handover? Are phased handovers required?
5.3. Are there any modifications to standard working hours? Will weekend/night working be permitted?

Communication
6.1. Lines of communication should be established at the onset of a project: these should include, as a minimum, the names of persons within each organisation who will be responsible for the work to be undertaken, and the names and addresses or telephone numbers of the persons who will have responsibility for the day-to-day running of the job (if they are different)
6.2. As early as possible a series of meetings should be established which will allow discussion with the various parties involved in the project
6.3. Methods of communicating progress/cost to the client should be agreed

effective way to carry out the work — is probably the client's most important decision. The range and combination of professional and commercial relationships which are possible are countless and can so easily be wrong. It is the author's firm conviction that these relationships can only be assessed properly by a system of second-party auditing which begins for the client with the project manager and goes down to the last subcontractor's appointment. In establishing the organisation for a project, all the problems of communication, motivation, competence and skill, quality standards and availability of resources cannot be ignored, and these and many other factors require expert assessment from the outset.

Implications of the brief

The process of second-party auditing, even if carried out by a third party on the client's behalf, for a specific building project is much concerned with the relationships between the building owner and the principal parties. What is required from these relationships will depend on the knowledge of the building owner and the contractual framework he intends to establish. The effectiveness of the process will depend largely on the quality of communication. The client must initiate that communication, and make the appropriate decisions as a result of it.

In order that all the parties fully understand their likely commitment, it will be necessary to brief them at the outset on the nature and scope of the prospective project. This process of informing has to start with the client briefing his design team. In turn they must be able to give the main contractors and thus the subcontractors a clear understanding of the work requirements and any constraints. This information may at the outset be conceptual in nature, but it will be built on by the design team before being presented to the contractors and subcontractors at a later stage.

Care should be taken in providing a briefing document, which will of necessity require enlargement and modification as the project proceeds. (How this is done will depend on the nature of the particular project.) The briefing document should also be used to give salient information to the auditees so that they are able more fully to respond to the prospective audit.

Scope of project

The invitation to participate in the project and the request to allow a second-party audit to be carried out before the contract is entered should contain the information shown in Table 19.

Clearly the level of information that is given will be much greater for the later audits than can be given to the project manager and various members of the design team, who will themselves fill out the details for later audits of contractors. But this is where project quality management begins.

So that a client may proceed with this all-important stage himself, appendix 3 comprises a model document for carrying out a second-party audit on the design team. Similar questionnaires are used for general contractors and for specialist contractors, and by general contractors for domestic subcontractors.

The first approach to the design team will probably begin as a series of theoretical aims from the client, which will have to be translated into more practical parameters by the design team.

Auditor's first visit

Appendix 4 gives further guidance notes for an auditor on his first visit to a company in carrying out a second-party audit.

Having given the information indicated to the auditee the second-party project audit questionnaire (Appendix 3) should then advise him of the areas that the second-party auditor will wish to cover on the audit visit, and so enable the prospective contractor to answer in a meaningful manner.

12 Auditing works in progress

Whether total quality management has been exercised from the conception of a project, or is only instigated by the contractor for his own benefit, the auditing of subcontractors and of the site construction activity will still be necessary. The issue of line management authority or policing by a quality manager will have to be resolved.

I T CAN BE assumed that by the time works have begun, or are about to begin, on site, the information given to the general contractor and principal specialist contractors has provided the necessary framework for them to respond and develop their own project plans. Their quality systems will have been subjected to an appropriate pre-contract audit, and the total management plan — or rather its main parameters and key events — will have been provided.

As part of the main contractor's development of his management plan there should be the inclusion of appropriate pre-subcontract audits. Here the contractors who have regularly used particular subcontractors' packages may already have records of their past performance and so the project-specific audit will need only concern itself with matters related to differences on this particular contract. There *will* always be differences, and skipping or omitting this audit altogether will not be advisable. Certainly the existence of a third-party certificate in general terms should never be taken as sufficient evidence to justify the suitability of any subcontractor to carry out a particular project.

There remains then the audit during the progress of the works.

In some cases the project may have come to the starting gate without having gone through some or all of the steps of good practice outlined in the preceeding chapters. Thus the first audit on site after works have started should go through

Table 20 (facing page). Construction project progress audit

Preliminary
A.1. Is there a project quality management system?
A.2. Is there an organisation structure?
A.3. Has it been distributed to all who need it, in a project manual?
A.4. What is the distribution of this manual?
A.5. Is there a quality manager?
A.6. Who does he report to? What for?

Construction phase
 1. Ascertain if a firm date has been agreed for start on site and if this is acceptable to all parties
 2. Check if comprehensive insurance has been arranged for all risks, including possible damage to third parties and adjacent premises
 3. Check if the contractor's detailed programme has been received and if it is acceptable
 4. Check if a progress monitoring system has been set up using method such as time/cost graphs, critical path analysis, computer network planning or target completion dates
 5. Ascertain if the contractor has provided 'information required' schedules and if there are copies of relevant sections with each consultant
 6. Check if a system of monitoring information-required schedules is in hand
 7. Check that all consultants are providing the contractor with adequate information and that they are doing so by the dates required
 8. Check that all consultants are providing adequate site monitoring
 9. Ascertain if communication with the client is regular and if facilities exist for him to attend site
10. Ensure that project meetings and design meetings are continuing
11. Ascertain if site progress meetings are run regularly and formally, with minutes taken and circulated
12. Ensure that site inspections are made frequently, both with and without members of the design team
13. Ascertain if architect's instructions/variations are copied to all members of the design team
14. Ascertain if there would be an agreed maximum value of instructions/variations above which the client must be first consulted
15. Check if valuations are prepared quickly
16. Check if architect is notifying the quantity surveyor of defective work to be excluded from valuations
17. Ascertain if the value of variations is being agreed quickly and as the work proceeds
18. Ensure that cost forecasts are produced regularly and on the basis of up-to-date information
19. Check if certificates are issued on time
20. Check if payment is made within the prescribed period
21. Check if there are any tell-tale signs that the contractor may be in financial difficulty
22. Ascertain if delays are occurring in the contractor's programme; check if extensions of time are in order and what the cost/delay implications are to the client
23. Ascertain if it is appropriate to recommend deduction of liquidated damages
24. Ascertain if the quantity surveyor and contractor are maintaining the momentum of the contract period in settling variations and agreeing the draft final account
25. Ascertain if a date has been fixed for presentation of the draft final account
26. Decide if discussions are necessary regarding effect of loss and expense claims and liquidated damages
27. Ascertain if delays are being caused by nominated subcontractors
28. Decide if further cost forecasts should be prepared during final account stage
29. Ascertain if the contractor's attention is being drawn to major defect items which arise during the defects liability period

checking procedures to see that all the planning work has been consolidated with the site manager.

Whether done by or for the client, the audit should cover the checklist items in Table 20. If the contractor is using a quality manager separate from the line management, his list will probably be very similar.

It is the firm view of the author and all his colleagues that, in construction projects, quality can only be built in, not checked out. Quality should therefore be a line-management responsibility for the general contractor and for each subcontractor. The traditional monitoring carried out by architects, surveyors and consultant engineers should provide sufficient external monitoring save for the rather special third-party adjudication required for construction projects, which is dealt with in chapter 13.

13 Contract management adjudication: the final step in project quality management

It is implicit in total quality management that traditional attitudes that have prevailed in the management of construction companies and projects must change. In construction the teamwork for total quality management will not survive the difficulties experienced in construction on site without a means of continuing the harmony between parties in a conflict situation when these problems arise. Contract management adjudication is a new concept developed to continue pre-site harmony in the heat of construction endeavours. Management techniques must provide the procedures and practices as tools to give effect to the principles, which evolve, essentially, from a philosophy. These tools are set out as the means of retaining harmony without prejudice to any of the parties' ultimate legal rights of redress. The chapter shows how they will add value and reduce the cost of quality by eliminating or subsequently reducing the costs of dispute and subtracting the negative value of legal fees.

THROUGHOUT this book the emphasis is on the quality management techniques, philosophies and disciplines which are aimed at improving project performance, and therefore profit to all parties — contractor and subcontractor, design team and client.

For the management plan to be fully effective, the project quality philosophy, principles and procedures must start at the top — and at the beginning of the project process — and thus with the client, or his representative and project manager. It is assumed that this has been done, the necessary goodwill established, and an effective team spirit developed, aimed at the achievement of the common objective, the industry's common goal of customer satisfaction — the achievement of the client's requirements in terms of function, aesthetics, cost and time — a quality building. This can only be achieved in practice on site, where the unexpected can be expected. So, frequently, changes must be made not only to the project's construction quality plan but perhaps also to the detail design, or even the requirements for the design. It is for this reason that 'claims' arise far beyond those provided for in provisional or pre-contract sums.

This is the stage when claims rapidly lead to disputes, for the protection and self-interest of the many parties involved. The team spirit can then be quickly shattered on subterranean rock!

Contract management adjudication

But the team spirit must prevail if the project is to be successful. This is the place for a new management technique. It is akin to auditing and has a similar purpose. From a combination of the philosophies of total quality management and arbitration there has evolved the concept of contract management adjudication.

Total quality management begins, as shown in earlier chapters, not just at the construction phase, but at the time the client's brief is developed and passed to the architect. Then, the use of quality management procedures will have eliminated many of the communication problems that can beset a project, frequently delay its completion, increase its costs and lead so often to the depressing pattern of claim, counterclaim, dispute and ultimately litigation, or arbitration, which have in the past beset the industry with benefit to no-one except the lawyers.

Cost of construction disputes (quality costs)

The value of construction disputes being resolved in 1992 was reliably claimed to be £3000 million, amounting to some 7% of the total value of all construction output in 1992. An analysis in *New Builder* of 10 September 1992 of disputes in the Official Referees' Courts showed that almost a quarter of the 755 writs lodged were in construction, and indicated that £436 million would be expended in pursuing disputes through the courts by the end of 1992. Typical legal costs would be a further 25% of this figure. But this in many ways is only the tip of the iceberg: arbitration, mediation, alternative dispute resolution and commercial battles all add to these costs.

These extremely high costs, which are of poor quality in the overall construction process, will be eliminated by ensuring that quality management systems are used on the project from start to finish.

Total quality management systems will have improved performance, reduced costs and increased profitability thus far. But because all construction projects have a built-in potential for dispute through conflicts of interest across individual one-off contracts, there is need for a further process to keep the industry harmonious and 'out of the courts'. This in itself will also have a substantial effect on management stress, company morale, public relations, client relations, profitability and cost.

Multidisciplinary projects have grown larger, more complex and more prone to confusion and conflict between the parties. Legal dispute has been the growth industry of the past decade.

Good management, developed through the total quality management process, is undoubtedly the preventive medicine of dispute, and it will reduce the costs of poor quality in the development process. But differences will still arise on site, and good management of the potential dispute situation then becomes an even more vital requirement.

Third-party adjudication

Starting with total quality management, alternative procedures to resolve technical problems in construction are long overdue. This is where techniques involving partnering and adjudication can be of value. These are concerned with utilising the benefits of arbitration, recognising that the technical content of a dispute is at least as important as the contract clauses under which the project has developed, and with giving an adjudicator (or auditor) greater freedom to intervene, when asked, than is customary in English arbitration.

The Adjudicator's role is more akin to that of a cricket umpire than that of a supervising architect or engineer. Normally he intervenes only when one of the sides appeals for a decision as to whether a batsman is out, but he also has other duties to observe 'no-balls' and offences against the rules that might otherwise go unobserved or cause dispute. Adjudication techniques are equally applicable to contractual disputes concerning materials supplied, or work done, and are particularly suited to claims where an insurance company is also involved indemnifying and so backing one (or sometimes both or several) of the parties.

All aspects of the conduct of the project should be covered by adjudication. This is possible if these provisions have been built into the project's management plan before the documents go out to tender to both main and subcontractors.

The tendency within the construction culture has been to regard litigation, arbitration and adjudication as necessary evils to be invoked in order to settle disputes which are already under way. There is, however, a better role for adjudication and that is as an extension to the total quality management of a contract. This is a powerful management tool which should be used early, positively and dynamically to improve performance in an area which currently wastes that most important resource — management time.

Adjudication is cheaper and quicker than other means of settling disputes or claims. Better still, as the last stage of total quality management, it reduces the chances of unresolved dispute ever occurring. Contract management adjudication as a method of resolution develops the advantages of third-party auditing into arbitration for technical disputes — where the arbitrator is chosen for his knowledge of the matters in dispute and can let his expertise influence his judgement. As a monitor of the project-specific management system, he will be readily on hand and able to provide instant umpiring, not just reporting via a neutral communication channel to one or both parties.

Benefits of adjudication during progress of the works

The problems of litigation or arbitration on technical issues are as follows.

- The action is normally brought long after the event, when witnesses' memories have dimmed.
- The case often turns on expert opinion as to the state of the art several years previously.
- The vast quantity of documentation takes a great deal of

time for everyone concerned with the action to read and absorb, adding greatly to the cost.

- The English adversarial system of litigation necessitates solicitors, advocates and experts being retained by both parties, or, in the not infrequent case of seven or eight defendants, by all parties to the action; all adding to the time and cost of reaching a conclusion.
- Sometimes, too, the legal outcome is one on which the judge and the lawyers agree, but the parties and their technical experts have great difficulty in reconciling with their knowledge of the technical facts.

Arbitration can, and frequently does, provide a better resolution than litigation for technical disputes. Contract management adjudication advances the time of resolution to the time of the incident and greatly reduces the quality costs.

Recent years have seen several developments of adjudication technique. The British Property Federation has formulated a project management system in which the role of an adjudicator is defined. In the USA a system of adjudication has been developed to resolve disputes by 'mini-trials' and 'pre-trial reviews', where the main issues are examined by a judge and lawyers and technical experts, but in an informal hearing of predetermined and limited duration (and therefore cost). All these techniques have the objective of a quicker, cheaper and more business-like resolution of the dispute.

However, greater benefit can be obtained on long-term or complex projects if the adjudication principle is established by the parties at the outset as part of the monitoring and auditing process of the contract, and this should be built into the total quality management plan. Observations can then be made by the auditor on the contract documents and on preventive measures throughout the plan's operation. This is the technique developed by Polycon consultants in the early 1980s and known as contract management adjudication (Fig. 21).

Purpose and procedures of contract management adjudication

The purpose of contract management adjudication is to improve performance and reduce costs to all parties

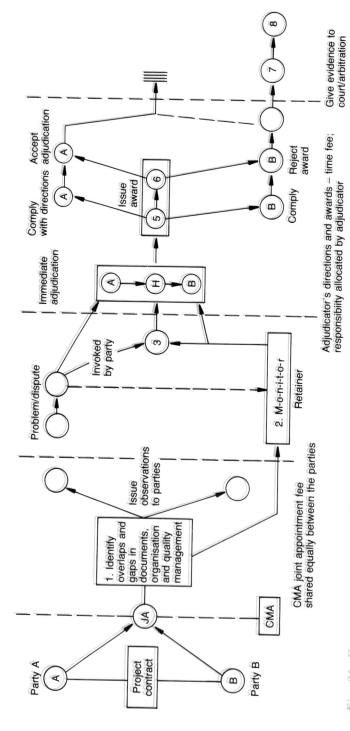

Fig. 21. Contract management adjudication (CMA) during long-term contracts: outline procedures

throughout large or multidisciplinary projects. It recognises that disputes are sometimes inevitable and that their speedy and economic resolution will be made by objective and impartial decisions from experienced professionals who have not been involved in the design, development or formulation of the contract. It also recognises that the authority of professionals and contract management within the project's total quality management plan must not be disturbed.

At the start of the contract, the adjudicators examine the project to identify gaps and overlaps in the contract documents, the contract organisation and (to the extent that it exists already) the project management system.

Traditionally, forms of contract have often named the architect or engineer as quasi-arbitrator. Yet they already have a contractual relationship with one of the parties. Further, likely subject areas for dispute result from their design or organisational decisions. A designer may have omitted data from his specification, on which contractors based contractual promises; or the designer may have made what turn out to be wrong design or procedural decisions, costly for the contractor to correct.

Problems occur most frequently on complex projects if the lines of communication and lines of contract are not co-ordinated at the onset. An initial appraisal is carried out by examining the project to identify gaps and overlaps in the contract documents, the contract organisation and the building contractor's project quality management plan to ensure that this co-ordination is in place.

The project's quality system will not be fully effective unless the activities of the client and the design team have been properly integrated, as they should have been where total quality management has been involved. After this initial appraisal, or project audit, the adjudicator should make his observations to the parties — objectively, factually and in writing.

Thereafter, a watching brief is maintained by the auditor/adjudicator. When any retaining party calls for a ruling, the adjudicator or his link man will decide whether he can deal with the matter himself or whether he needs to call in others with appropriate technical or legal background to act jointly with him.

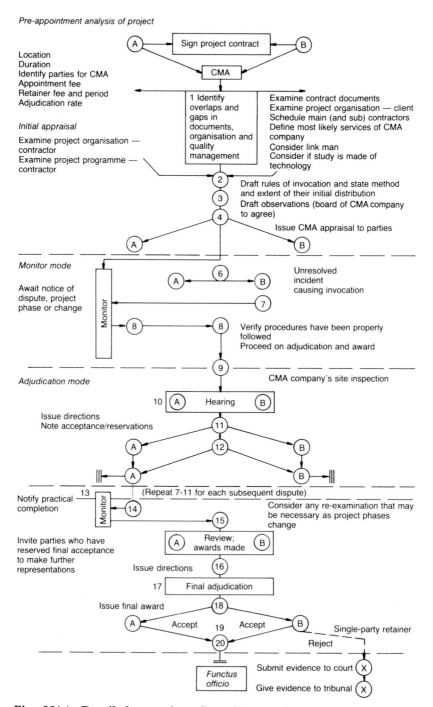

Fig. 22(a). Detailed procedure flow diagram for contract management adjudication (CMA)

PRELIMINARY

A. The appointment of Polycon AIMS Ltd. as Contract Management Adjudicator can be made at any time after the project contract has been made between the parties. It may also be signed after an incident has already occurred causing a dispute during the continuation of the contract.

B. As a first step in either situation the CMA Appointment Form must be signed by both parties - then the operations will be carried out by Polycon AIMS through a "Link-man" who will be delegated with the authority as Adjudicator and/or Chairman of Tribunal of other Polycon AIMS specialists whose experience may be required by the subject-areas of any dispute.

INITIAL APPRAISAL

1. A Link-man will be appointed by Polycon AIMS's Board and notified to the parties. The Link-man will then seek or receive from the parties the main project contract documents, the project organisation documents, if any, that apply to both client and the main contractor's organisation, the project programme or network, a schedule of the main and sub-contractors already nominated and involved. The Link-man/Adjudicator will examine these documents to identify gaps and overlaps in:-
   a)   the contract document;   b)   the contract organisation; and also consider whether any study should be made of the contract technology.

2. The Link-man/Adjudicator will then consider whether any other Polycon AIMS member should be nominated to the project tribunal and whether anyone else should also study the contract documents.

3. He will then draft the invocation rules and state method and extent of the initial distribution of these rules. He will also consider whether any other contractors or sub-contractors, agents, etc. should be covered by the Rules for CMA Procedures. He will draft his observations, submit and agree these and the Invocation Rules with Polycon AIMS's Board.

4. Polycon AIMS will issue the CMA Appraisal and Invocation Rules to the parties.

MONITOR MODE

5. The Link-man will ensure that the parties have received and understood the rules for invocation and will then await notice of any dispute.

6. Upon the incidence of any matter causing a dispute to arise either party invokes the procedure rules and gives notice of a dispute requiring Adjudication.

7. The Link-man verifies that the Invocation Procedures have been properly followed and then arranges to carry out an appropriate investigation/site inspection.

ADJUDICATION

8. The Link-man visits the site (accompanied by Adjudicator or Specialist?)

9. Link-man/Tribunal holds meeting(s) on site and hears representations from all parties involved.

10. The Link-man issues Directions and any interim award to both parties.

11. The Link-man notes acceptance of his Award or any reservations of the parties.

12. Return to **Monitor Mode** to await further incidents, notification of change of main contract phases, or Practical Completion.

13. The parties notify Practical Completion of the contract and indicate whether they consider a further Adjudication or Final Award is required.

14. Link-man and/or Tribunal reviews interim awards, and if necessary;

15. Issues Directions for any final hearing prior to considering & drafting Final Award.

16. Link-man/Tribunal drafts final Adjudication Award & submits to Polycon AIMS's Executive Board - any dissentions or minority opinions are considered, and Award published.

17. Company Secretary ensures all fees are collected and issues Final Award to parties.

18. Company Secretary monitors acceptance (functus officio), or:

19. Company Secretary receives notification of rejection by one party and advises Link-man. Checks that all fees have been received.

X.1 Company Secretary awaits request by any party for copies of Award or attendance of Link-man or others to give evidence.

X.2 Company Secretary keeps Adjudicator/Witness informed of dates of Court/Arbitration hearing.

Fig. 22(b). Activity schedule for contract management adjudication (CMA) by Polycon AIMS Ltd ('the CMA company')

The purpose is to 'manage a potential dispute situation' so as to dynamically and economically resolve problems as early as possible in the chain of causation. Although, in principle, any skilled construction professional who is not a party to the project or to any contracts established within its framework could fulfil this role — given suitable training. In multidisciplinary projects a company possessing all the technical disciplines and trained in the principles of dispute resolution, able to respond technically no matter what the technical nature of the problem is, will be more effective in terms of cost and time. A company, Polycon AIMS Ltd, has been developed specifically for this purpose, and has already successfully resolved several disputes.

Figure 22 shows a detailed procedure flow diagram and activity schedule for contract management adjudication.

The advantages of technical adjudication by a contract management adjudication company are shown in Table 21. Reference to these advantages should be included in the total quality management plan and so in invitations to tender and other documents.

Contract management adjudication, used as a project management technique and part of the project auditing process, will

- reduce avoidable delays during the progress of the contract
- prevent dispute immediately a situation arises that might damage one of the parties or prevent it from proceeding with the work
- reduce the costs to all parties involved in the dispute
- eliminate the 'judge in own cause' situation and conflict of interest of the professionals whose earlier decisions become an issue
- document factually and objectively all issues dealt with, and give expert and impartial opinion should any party still wish to contest them later and in court
- monitor and audit the project quality management system.

The process thus combines concepts of post-contractual arbitration, organisational project management, and impartial auditing of the project quality system.

Table 21.   Advantages of technical adjudication by a contract management adjudication company

---

1. It is the company that makes the Award.

2. The cost of obtaining the Award is far less than with litigation or arbitration, and can be predetermined.

3. All aspects of a complex dispute are dealt with through the multidisciplinary composition of the company's expert and arbitrator members. These reflect the total range of disciplines which are involved in the conception, design, development and cost negotiation of a project — the contract — and its execution. If there is a particular area of (narrow) expertise required to investigate a particular point of conflict which is not contained within the company's expertise, a technical assessment is obtained from a neutral specialist — if necessary, the best on a world-wide basis.

4. This eliminates the multiple costs for sets of experts for each or all parties. Whatever expertise is required is contained in the adjudication tribunal and then is built into the Award.

5. The parties can reserve their options and so do not have to commit themselves to either final acceptance or rejection until the end of the contract. Then, if not objected to, after a predetermined period it becomes final and binding.

6. All Awards made by the company tribunals are scrutinised by other directors and countersigned before issue to ensure that the reference has been fully and properly discharged in accordance with the company's contract of joint appointment and that there are no errors on the face of the Awards, or in their preparation, and the conduct of the reference.

Contract management adjudication model clause

It is not necessary to make any major changes to the basic standard contract documents used in UK construction projects for the incorporation of either total quality management or contract management adjudication.

In the case of total quality management, it could be said that the culture and conventional contract documents have always assumed co-operative, effective, and well-motivated performance by all parties involved in the nexus of contracts. But what has happened, particularly over the past 10 – 15 years through the onset of excessive competition, is a confrontational situation of conflict between parties, spurred on by an ever greater use of litigation lawyers. This has also been at a time when there has been great demand for quick completion times and short-term and excessive profit, resulting in poor management of the overall project process.

As far as contract management adjudication is concerned,

all that is needed is a clause in the invitations to tender, and the inclusion of provision in the project specification, followed by a single clause in the contract documents as set out below.

The British Property Federation contract system and the Association of Consulting Architects form of contract incorporate clauses dealing with adjudication, but a service has been developed by Polycon AIMS Ltd which gives practical and highly cost-beneficial effect to the principles behind the clauses of the British Property Federation contract. It was developed separately but overcomes the disadvantages that exist with these and other adjudication techniques.

Polycon AIMS's adjudication and arbitration procedures stem, as does all arbitration, from the authority of the parties. They embrace concepts of the powers given to arbitrators under English law and some of the more normal practices which flow from Continental legal practices.

Contract management adjudication in particular has been developed to be applicable to, and usable with, any and all of the standard forms of construction contract. It does not require any change to be made to the standard forms and thus can be implemented immediately, but it is obviously better to build it fully into the project quality plan from the commencement.

Contract management adjudicators can be appointed at any time after the main contract has been signed, whether or not a difference has already arisen. But it is better that provision should be made in the specification prior to the signing of the main contract, in which case the following clause should be included.

> In addition to/substitution for the provision for Arbitration under Article 4 of the Recitals, Contract Management Adjudication procedures shall apply to disputes or differences arising out of or in the course of the Works.
>
> It is also intended that . . . of . . . shall be appointed jointly by the Employer and Contractor to resolve any dispute or difference that may arise during the progress of the Works between the Employer or his Architect, Engineers, Quantity Surveyor or Supervising Officer and the Contractor or any Sub-Contractors, to issue Directions for the continuance of the Works under these conditions but without prejudice to the right of any party to seek

formal Arbitration if they do not accept the Adjudicator's Award as final and binding upon them all at the end of the Contract.

The Contractor will be required to incorporate the provision for Adjudication in any sub-contract for the Works and for the supply of items for the Works whether these are from nominated persons or otherwise.

Advantages of technical adjudication by a company

Few firms are yet in a position to ensure that a total quality management plan and organisation system will be implemented throughout a project's life by a complete team, each part of which is committed to total quality management in its own sphere of activity. However, even if this is possible there will still be the need for effective means of maintaining the motivation for, and commitment to, the project — rather than the individual interests of various parties involved in the nexus of contracts that has been created — when post-contractual conflicts of interest arise.

Contract management adjudication can play an important part in reducing the effects of these conflicts and should be seen as the final link in the total quality management chain.

14 Conclusions and action plan for quality projects

WHAT THEN are the conclusions that can be drawn from the analysis made in the preceding chapters?

First, total quality management — the third industrial revolution — and its principles apply to the construction industry.

Next, the total quality management philosophy of teamwork and co-operation, not confrontation and conflict, is long overdue for the construction industry and on construction projects. This does not mean the end of competition, although it should certainly mean the end of blind acceptance of the lowest tender unless the assurance of quality management has been established by pre-contract audit.

Third, the concept of gradual and unending improvement in performance is as essential to the construction industry as it is in developing countries and their economies and in the manufacturing industries of sophisticated Western nations.

Fourth, education and training in teamwork at all levels and in all areas is essential for construction. In the construction professions particularly and in most of the companies making up the industry on site, this means management training for managers. This is of even more importance than craft and skills training at operative level. Management must be recognised as a discipline separate from the technical and professional skills.

Fifth, to stimulate and maintain this unending improvement in team performance, monitoring will be required through quality auditing. The concepts of first-, second- and third-party auditing will be involved, and in that order of importance. Management must first satisfy itself that its own organisation is performing well and reliably, and also seek permanent improvement, which it can then demonstrate to any second-party auditor. The essentially project-related nature of the industry means that second-party auditing must be project-specific.

Sixth, any third-party auditing must relate to the project situation and to the team assembled for the project. The construction industry's product is one-off and therefore the assessment must be made against the *project's* defined standards if it is to be fully effective.

Seventh, the concept of third-party auditing, if it is to add value and not cost to the construction process, must involve judgemental qualities and a direct relationship to, but not involvement with, the project and project team. The game can only proceed if the third party provides, as well as the monitoring role, elements of umpiring similar to those exercised by referees of organised competitive sports.

Eighth, progress within individual construction organisations must be through management and leadership to achieve commitment to, and through, a corporate philosophy of customer satisfaction that recognises that business goals are inseparable from customer satisfaction.

Ninth, companies must document procedures and systems to achieve high standards and continuous improvement through permanent and comprehensive training systems.

Finally, organisations must see that they produce the continuous improvement required and then monitor these improvements by regular and internal value-added audits.

If this book appears to deal more with the philosophy and principles of total quality management and with the reasons for the auditing procedures rather than with the techniques of management auditing itself, this is because only too often, not least in the construction industry, manager and managed (workers all) are trained to carry out tasks without education into why and where they fit within the overall philosophy and pattern of teamwork within the corporate structure.

In construction this teamwork must have its emphasis on the client's project. The involvement of the customer in the achievement of his objectives within the contract must be closer than in any other industry's customer relationship and activity. Perhaps the only exception would be in the case of a medical team bringing a sick patient back to health, for here too, without the patient's motivation, help and participation, the best medical skills, technical knowledge and equipment will fail in their task.

It is through this knowledge that both job satisfaction and what Deming calls '*the joy of work*' will come, and so provide the motivation to improve both individual and corporate performance.

Clearly, the whole construction industry is project-orientated, so improved quality performance must be project-related and include the whole project team. The manufacturer, the specialist subcontractor, the main contractor, the professional designers, the project managers and, above all, the client — the customer the whole industry exists to serve — must be involved in the process.

Quality auditing has an essential part to play in improving performance in construction practices and on projects through total quality management.

APPENDIX 1
The European Quality Award

The following is based on extracts from the award literature to illustrate the importance given to self-assessment in developing total quality management.

THE EUROPEAN Foundation for Quality Management (EFQM) has a membership of nearly 200 leading European businesses, all of which recognise the role of quality in achieving competitive advantage. The EFQM is committed to promoting quality as the fundamental process for continuous improvement within a business.

The European Quality Award incorporates the European Quality Prize and the European Quality Award. The European Quality Prize is awarded to a number of companies that demonstrate excellence in the management of quality as their fundamental process for continuous improvement. The European Quality Award is awarded to the most successful exponent of total quality management in Western Europe. The trophy is engraved and held nominally for one year by the recipient.*

The first step in a company's application for the European Quality Award is the collation of a body of quality management data from within the organisation. There is significant value in this process, even if the company is not successful in winning the award. It enables those involved to assess their own company's level of commitment to quality. It also shows them the extent to which that commitment is being deployed: vertically, through every level of the organisation, and horizontally, in all areas of activity.

The award assessment model looks at people, processes and results. Processes are the means by which the company

---

* The first European Quality Award, made in 1992, was given to Rank Xerox Ltd who, to quote their chief executive, 'were brought back from the brink of extinction in 1989 by introducing a programme of total quality management'. Winners of the European Quality Prize were BOC Special Gases, Milliken European Division and Industrias del Ubierna SA, UBISA, of Spain.

harnesses and releases the talents of its people to produce results. Thus the processes and the people are the enablers which provide the results. Fig. 23 shows that customer satisfaction, people (employee) satisfaction and impact on society are achieved through leadership driving policy and strategy, people management, resources and processes, leading ultimately to excellence in business results.

The nine elements shown in the model correspond to the criteria which are used to assess a company's progress towards excellence. The results criteria are concerned with what the company has achieved and is achieving. The enablers criteria are concerned with how the results are being achieved.

For the purposes of meaningful assessment for the award, relative values must be ascribed to the nine criteria within the model. These are obtained by consultation involving EFQM members and many other European institutions.

The application process for the award requires that a company presents its own performance across a range of specific areas relating to each criterion. The criteria are written in non-prescriptive terms, which allows companies freedom to put self-appraisal information which is relevant to their business situation.

The requirements for the enablers criteria are as follows.

- Information required on how the company approaches each criterion. Each criterion is covered by a range of specific areas and the application should provide concise and factual information about each of these areas.

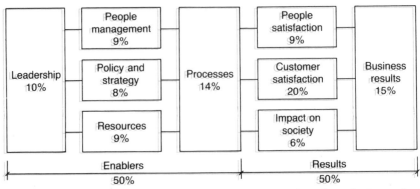

Fig. 23. **Award assessment model: the percentages are the relative values (as used for the 1992 award)**

- Information is required on the extend to which the approach has been deployed — vertically through all levels of the organisation, and horizontally to all areas and activities.

The following are the requirements for the results criteria.

- Information is required on the key parameters the company uses to measure results.
- For each key parameter, data are required, ideally in the form of trends over three or more years. The trends should highlight

  — the company's own targets
  — the relevance of the parameters to all groups with an interest in the company
  — the company's actual performance
  — where appropriate, the performance of competitors and/or the performance of 'best in class' organisations.

- Information is required on the extent to which the parameters presented reasonably cover the range of the company's activities. The scope of the results is an important consideration for the assessors.

It is anticipated that the UK National Award run by the British Quality Foundation will follow similar lines and use similar criteria.

APPENDIX 2
Guide to quality systems auditing
(BS 7229, 1989)

B S 7229 was prepared under the direction of the Quality, Management and Statistics Standards Policy Committee. It was developed in conjunction with an internal draft on quality auditing, planned to be titled 'Generic guideline for auditing quality systems', in the work programme of technical committee ISO/TC 176, Quality Management and Quality Assurance. It was considered by the British Standards Institution, however, that early publication as a British Standard guide was desirable in view of the increase in audit-related activity in the UK.

The introduction states that the guidance is applicable equally to any one of the following three specific and different auditing activities.

First, a quality system audit can be carried out by a company on its own systems for the purpose of giving assurance to the management that its quality systems are effectively achieving the planned quality objectives. These audits, known as first-party, or internal, audits, can be carried out either by the organisation's own staff, provided that they are independent of the system being audited, or by an outside agency.

Secondly, an audit can be carried out by one organisation on another, with whom they have, or intend to have, a contract to purchase goods or services. The purpose of this audit is to provide assurance to the purchasing organisation that the quality systems of the supplier are capable of sustaining the delivery of products or services, if already under contract, or of providing those goods or services to an agreed quality level. These second-party audits can be carried out by the purchasing organisation or by an outside agency.

Thirdly, third-party audits can be carried out by

independent agencies that may be accredited, using a national or international standard such as parts 1 – 3 of BS 5750, (ISO 9001 – 9003) to provide assurance on the effectiveness of the quality systems.

In the case of a third-party audit the purpose of the audit is to obtain confirmation of compliance with the standard. This can be used to provide assurance to existing and prospective customers for the product or service.

The standard establishes basic auditing principles, criteria and practices, and provides guidelines for establishing, planning, carrying out and documenting quality systems audits.

Definitions

The term 'assessment' in the standard is accepted as being the same as and interchangeable with the term 'audit'. Section 2 of the standard defines terms as given in ISO 8402 (BS 4778).

Objectives and responsibilities

Section 3 deals first with the general and specific objectives for the audits. It then deals with *incorrect objectives*: these are such that result in a transfer of the responsibility from operating staff to the auditing organisation, or lead to an increase in the scope of quality functions over and above that necessary to meet quality objectives.

The section goes on to deal with the auditor's responsibilities for

- complying with applicable audit requirements and standards
- planning and implementing the responsibilities effectively and efficiently
- reporting results
- security of documentation and information

  ensuring confidentiality
  treating privileged information with discretion

- maintenance of independence.

The additional responsibilities of the lead auditor are also covered.

The section deals also with the minimum requirements for the competence of the auditor and of the audit team; and with the roles of client and auditee.

Auditing

Section 4 deals with the scope and frequency of the audit, preparation for the audit, the audit programme and the working documents required, execution of the audit (covering the opening meeting, obtaining the evidence and the closing meeting with the auditee), and the audit report, its content and its distribution.

Audit completion

Section 5 deals with the completion of the audit and the retention of audit documents.

Corrective action and follow-up

Section 6 covers the corrective action that should be taken by the auditee, and states that the auditee should notify the client of the corrective action taken so that the client may verify by audit if necessary and report the outcome.

APPENDIX 3
Quality management audit: second-party project audit questionnaire for client/design team professionals

The following pages show a facsimile of the second-party project audit questionnaire produced by TQM Polycon Ltd.

PRELIMINARY INFORMATION

2PQ NO. ....

(to be retained by Audited Company)

Client Name:	
Address:	
	Tel: Fax:
Project Title & Reference	
Client's Project Manager/ Representative	
Nature of Project (attach any documents that you think will assist)	
Intended Timing of Project's Main Phases	
Likely Order of Cost:	
Total Building Cost	£
Any Phases:	£
Likely order of cost of the particular package of work that this Audit covers	£
Preferred timing of this work package	
Date for return of this Questionniare	/ /
Date for placing Letter of Intent (Contract/Order)	/ /
Dates for Commencement of work (covered by this package)	/ /

Module D.212.A
11/90

QUALITY MANAGEMENT AUDITING
2nd Party Project Audit

CLIENT / DESIGN TEAM PROFESSIONALS

1. The purpose of this Project Questionnaire is to establish the comprehensive effectiveness of your management methods, systems and people to 'GET IT RIGHT FIRST TIME' economically and effectively. It will also have regard to your company's contracting and legal procedures with particular emphasis on measures to prevent contractual disputes arising.

   This audit is concerned with the totality of the factors that affect your company's Quality Management and could therefore be said to be of its 'Quality Management System', *ie: the organisation structure, responsibilities, activities & resources and events that, together, provide organised procedures and methods of implementation to ensure the capability of your firm to meet our specific project quality requirements.*

   Our aim is to develop a 'Project Quality Plan' which is a documented set of activities, responsibilities and events serving to implement its quality system.

   The Quality Plan must of necessity embrace all those involved in the project from the time of commencement. It will therefore include the work of the design team and their interface with Contractor and Sub-Contractor. It must also embrace the activities, operation and decisions made by or on behalf of 'the Client'.

2. This 2nd Party Audit, whether carried out by a Polycon consultant or by a project manager, is aimed at:-

   a) identifying the objectives and scope of the systems to be established;

   b) determining the present stage of development of your quality procedures and their documentation;

  c) providing a report for discussion by the quality policy or steering group, which will also provide the basis for the project quality programme leading up to final assessment / selection.

3. This questionnaire is intended to help in the selection of either professional organisations, or a Contractor / Sub-Contractor, through whom the Client wishes to carry out development or building works. This audit will always be limited to those aspects of the audited company relevant to the Client's specific project needs.

4. Many matters involved in the project quality system will affect more than one department or area of management responsibility and this questionnaire is intended to provide information that will aid selection. It is likely that the person receiving this may not be familiar with all aspects of his company, and so the questionnaire will form a good checklist for him also.

  This questionnaire is therefore divided into sections which may have some elements of overlap where activities are primarily the responsibility of one function and of the management related to this function.

  A. General Organisation - which should normally be completed by the Managing Director, Senior Partner, General Manager or Company Secretary.

  The following sections, where appropriate in a larger organisation, should be completed by the appropriate Departmental Manager.

  B. Marketing.

  C. Personnel, Industrial Relations and Safety

  D. Resource Planning & Finance

  E. Defects and Potential Claims

  Additional general questions related to particular Project and Client Company activities.

QUALITY MANAGEMENT AUDIT

A.   GENERAL ORGANISATION

1.   Name of organisation and address

Contact: Tel No: Fax No:

2.   Is this part of a larger group?   YES/NO
     Or are there subsidiary companies/
     locations?   YES/NO

     If yes, name & address of group
     and/or subsidiaries.

3.   What is the approximate number of
     employees directly to be involved in   0 - 45   50 - 99   Over 100
     the project which is the subject of
     this Audit?

4.   Please attach a list of current
     projects, locations and the approx.
     duration and contract value of each,
     or give an indication of the number,
     range and nature of the work your
     Company does and currently has on
     hand to convey a general impression
     of your current activities.

5.   Do you have an organisation chart?   YES/NO

     Please attach a copy, together with
     a typical Management Responsibility
     Mandate or Job Description.

6.   Does the Company have a Legal
     Adviser on the staff?   YES/NO

     If so, please give name and title.

7.   When was your firm founded?

8.   Is the founder still in the business?

9.   How many Partners/Directors are there?

10. How often do they meet regularly to
     conduct the overall business of the firm?

11. Who is / will be leading this specific
     project and what is the 'chain of
     command'?

QUALITY MANAGEMENT AUDIT

ORGANISATION PROCEDURES

12.a Is there an Office / Operational Management Manual?	YES/NO
.b What are its main sections?	
.c How is it referenced?	
.d How often is it updated?	
13.a Is there a Job Costing system?	YES/NO
.b Are Job Budgets prepared?	YES/NO
.c Who does / does not keep Timesheets?	
14. Are annual / quarterly / monthly Budgets prepared?	
15. Is a Drawing Register kept for each project?	YES/NO
16.a Is there a Technical Standards Manual?	YES/NO
.b What are its principal sections? (Please attach index)	
.c How, and how often, is it updated?	
17.a Does your Company have a computer?	YES/NO
.b If yes, will it be used on this project? If so, please specify how and for what.	
18.a Is there a training policy for all staff - technical and administrative?	YES/NO
.b Is a Register of Staff Training maintained? By whom?	YES/NO

QUALITY MANAGEMENT AUDIT

B. MARKETING YOUR SERVICE
    LIABILITY & NEGLIGENCE MATTERS

1. Which of the following services do you provide? Please specify your Company's primary function(s) & any secondary activities by ringing [A] thus and underlining thus [H] <u>Town Planning</u>	[A] Project Management [B] Architectural Design [B] Structural Design [D] Building Services Mechanical & Water Services [E] Building Services - Electrical [F] Interior Design [G] Quantity Surveying [H] Town Planning [I] Landscape Design [J] Other (please specify)
2. Do any of these disciplines provide their service independently of the others - and so arrange their own conditions of contract?	YES/NO
3. Who is responsible for ensuring that your Conditions of Contract are kept up to date, and properly effected for each project? How?	
4. Do you have a "Quality Manager"? Who is he? When was the appointment made?	YES/NO
5. Do you have a brochure which describes your service(s)? How is this distributed to potential clients? (Please attach copy of any promotional material available)	YES/NO ATTACHED
6. Please list at least 3 of your recent projects which can be visited by the Auditor to form an impression of quality.	
7. Have you had any Negligence Claims made against you in the last 5 years? Are there any outstanding now?	YES/NO YES/NO
8. Do you have any system for noting and recording any events which could give rise to a claim against you? How long are such records kept?	YES/NO
9. Do you maintain any systematic follow-up of your performance after handover and final account? What? How?	YES/NO

QUALITY MANAGEMENT AUDIT

C.   PERSONNEL, INDUSTRIAL RELATIONS & SAFETY

1. Do your senior employees have individual Contracts of Service?

    Does this cover the person you would intend to manage this project?

    YES/NO

2. Do you have a Works Council or formal Management / Union, Management / Worker meetings? If so, how often?

    YES/NO

    WEEKLY   MONTHLY   OTHER PERIOD (Please state)

3. Do you have an Employee Handbook?

    YES/NO

4. Does it include Personnel Policy? detailed Conditions of Hours, Pay, Holidays, Pensions, Notice, Safety & Security Precautions?

    Please attach a copy.

5. Have you ever been accused of unfair dismissal by any employee? Or of Sexual or Racial Discrimination?

    YES/NO   When?

6. If yes, what was the outcome?

7. Do you have:-

    a. a Safety Officer?

    YES/NO

    b. a Safety Committee?

    YES/NO

        If yes, who are members?

    c. Does the Safety Officer regularly visit your sites? At what intervals?

    NO/YES   Every   -------- days / --------   weeks

8. Please attach a list of the personnel, together with their CVs, who will be responsible for this specific project.

QUALITY MANAGEMENT AUDIT

D.   RESOURCE PLANNING & FINANCE

---

1.   Do you prepare annual budgets?   YES/NO

      If YES, please attach a list of headings & " "   Budget Headings attached

---

2.   Do you maintain a regular Budget Control   YES/NO   Quarterly?   Monthly?
      by systematic reporting?
                                                                   Some other period?

---

3.   Do you produce any Resource Planning   YES/NO
      Statements for the Company overall?

      Will you for this specific project?   YES/NO

      How?

      Please attach a copy of any standard
      form or format.   Copy attached

QUALITY MANAGEMENT AUDIT

E.   DEFECTS AND POTENTIAL CLAIMS

1.   Do you have any of the following, if so, give details:-

    i)   Professional Indemnity Insurance - with ?   NO/YES

          Total cover - date of commencement   £   / /   Expiry   / /

   ii)   Commercial Legal Proceedings Insurance with?   NO/YES

          Does it cover:-

          a)   Contractual Disputes   NO/YES   Total Indemnity   £

          b)   Licence Disputes   NO/YES   £

          c)   Property Disputes   NO/YES   £

          d)   Patents, Copyrights, etc.   NO/YES   £

          e)   Data Protection   NO/YES   £

          f)   Employer's Protection   NO/YES   £

          g)   Warranty & Indemnity   NO/YES   £

          h)   Directors' & Officers' Liability   NO/YES   £

          Other:

          i)

          j)

          Are there any territorial limits?   NO/YES

  iii)   What "excess" or part of the claim do you
          carry yourself?

2.   Do you keep records of complaints made against
    you/your service - by phone, or verbally?   NO/YES

          - or in writing?   NO/YES

    Please attach list or procedural description   ATTACHED   YES / NO / TO FOLLOW

3.   Do you keep records of claims made and modified
    to insurers?   NO/YES

4.   Have you an established procedure for, and who deals
    with, claims or potential claims notified?

    Please attach copy   ATTACHED   YES / NO / TO FOLLOW
                                                  / TO BE DISCUSSED

5.   To whom are they reported?

6.   Is every potential claim situation notified
    to insurers?   YES/NO

QUALITY MANAGEMENT AUDIT

SCHEDULE OF DOCUMENTS REQUIRED

For the first consideration of your firm's activities it will help us to complete our Project Audit expeditiously if you will supply us with copies of any of the following documents which are in normal project use in your firm.

A.   **STANDARD DOCUMENTS**

1.	Normal Letterheading(s) / cont. sheets	Attached / to follow / not available
2.	Copy of Objectives Clause in "Memorandum & Articles" (if a company)	Attached / to follow / not available
3.	Appointment Conditions/Conds of Service	Attached / to follow / not available
4.	Employee Handbook	Attached / to follow / not available
5.	Trade Union Agreements	Attached / to follow / not available
6.	Safety & Security Orders	Attached / to follow / not available
7.	Contract Management Procedures - inc. construction method statement, planning & programming	Attached / to follow / not available
8.	Estimating & Tendering Procedure Forms	Attached / to follow / not available
9.	Contract & Production Surveying Procedures	Attached / to follow / not available
10.	Standard Contract Documents - Please state if you use standard RIBA/RICS/CEE forms, etc	Attached / to follow / not available
11.	Equal Opportunities - Sex/Racial Discrimination - Policy & Procedures	Attached / to follow / not available
12.	Any Standing Procedures for: Accident Investigation/ Prosecution Policy / Rights of Search	Attached / to follow / not available
13.	Operations or Technical Procedures Manual	Attached / to follow / not available
14.	Quality Manual	Attached / to follow / not available
15.	Any other pro formas which will be in use for the project	

B.   **UNIQUE DOCUMENTATION BEARING UPON DISPUTE RISK**

1.   Professional Indemnity Policy

2.   Legal Indemnity or Protection Policy

3.   Claims Notification Procedure

Most companies have policy and procedure documentation throughout their organisation. Some will be in a complete manual of procedures. This will vary with the size and type of company. In some small companies it may be quite limited.

Please provide what you have.

PROJECT QUALITY AUDIT

1.   Having completed this Questionnaire, which addresses many factors and areas of your firm's organisation in order to assess your suitability for this project, will you please indicate below dates when it will be possible for the Auditor(s) to visit your firm to discuss this Questionnaire, and any other factors he may deem necessary, with the appropriate members of your staff, who should include key staff responsible for the project.

2.   The Auditor may visit our offices:-

on _____ between _____ and _____

OR

on _____ between _____ and _____

OR

on _____ between _____ and _____

3.   Have you any other comments which may have a bearing on this Audit?

Signature: _____   Date: _____

Position: _____

Please return the completed Questionnaire to the Auditor using the "Private & Confidential" label attached.

HAVE YOU TAKEN A COPY?

APPENDIX 4
Format and check-lists for visiting and reporting by second-party auditor

The following pages show a facsimile of the format and checklists for audit visits and reports produced by TQM Polycon Ltd.

Format And Checklists For Visiting And Reporting (By 2nd Party) Auditor

Guidance To An Auditor On Going Into A Company For The First Time

---

Following receipt of the completed 2nd Party Project Audit Questionnaire, (2PAQ) this Audit Visit is the next step.

1.   The Purpose of the visit is to examine the systems of the Company with particular regard to the objectives laid down in the brief, and to good managerial practice. The office(s) to be visited must be established as that to be concerned with the specific project which is the subject of this Audit.

2.   Before the Auditor visits the Company he will have reviewed the replies to the Questionnaire (2PAQ). The Auditor must analyse these replies in relation to the overall Project brief and should have established the clients strategy for the specific project under review. Concern for quality must begin from the very inception of the project and continue to its final conclusion, - that is commissioning and occupation. The Auditors client should therefore be fully aware of the extent of his own likely involvement and of the implications of the time and cost of the Audit - the benefits from which will be obtained in the quality and lower cost of the project construction.

3.   The visit of an Auditor to a Company should start with discussions with a member of the Board or a Senior Partner together with the person(s) who filled in the Questionnaire.

4.   The reasons for the audit must be clearly established with the senior members of the audited Company. On the first occasion it is part of a selection process and the final report should make clear recommendations to the commissioning client. If more than one company is audited against a single requirement the relative advantages and disadvantage, strengths and weaknesses, of the audited companies should be tabulated in relation to the project so that the auditors client can make an informed decision.

Ultimately this decision may be made on a competitive quotation for price or time. The audit should have established, the reliability and quality factors. The format of the report will clearly vary depending upon whether the audit is carried out before, or after a comprehensive bid has been obtained. The Auditor should remember that unless specifically so instructed his job is **NOT** to make the selection but to assess the criteria to enable his client to do so.

5. It is necessary to confirm any impressions given in replies to the Questionnaire regarding work currently being undertaken and the resources committed to such work. Part of the Auditors's task will be to examine carefully examples of previous or current work undertaken by the Company to assess quality and other aspects relative to the project which is the subject of the Audit.

6. Throughout the Audit the distinction must be made between how the company as a whole is run and what procedures come into play for individual projects. The likely Management Structure for the project under review together with the c.vs. of the senior personnel should be clearly established, and any additional current workloads for those individuals noted.

7. In companies where a method statement and a programme have been submitted as part of the Questionnaire response, the Auditor must ensure through discussion, that he is satisfied with such statements and that they broadly concur with the clients brief particularly in terms of time and cost.

8. Examples of Document Control Methods for other individual projects should be examined. From the point of view of a Quality System the documentation should include actions to be taken to check that what is required has been done correctly. A Document Control System requires the establishment of a method of recording the outward and inward transmission of important information including the acknowledgement of its receipt, and of registers to monitor these.

9. Methods of cost control and 'reporting back' must be established. The Auditor must be satisfied that any cost limitations will be fully understood and that methods are in place for dealing with any deviations and for corrective action.

10. Generally all the information required must be gained on this one visit. Copies of all relevant printed forms, procedure notes or manuals should be seen and if possible acquired and brought away for reference when preparing the Report to the Client.

11.   Where manufacturing is involved as part of the Company's expertise and project requirement it will also be necessary to visit the manufacturing premises and undertake an Audit on similar lines to the foregoing, but in this case the twenty sections of ISO 9000 Part 1 or Part 2 will be relevant. If the audited company has third party certification for these activities the certificate should be inspected.

12.   Discussion will be necessary to establish what procedures the company has to maintain quality records and any procedures that exist for improving standards and reporting back to senior management. It is particularly important that any such procedures are able to be translated from and into each project and can operate effectively on sites.

13.   The conclusion of an initial Audit will be a report to the commissioning client with recommendations in line with his requirements which may be:

    (a)   to establish the company's suitability to become a member of a 'Design Team' for a specific project.

    (b)   to establish the company's suitability to act as 'Main Contractor' or 'Managing Contractor' for a specific project either prior or subsequent to tendering for the works.

    (c)   to establish the company's suitability to act as a sub-contractor for a specific project either prior or subsequent to tendering for the works.

In the latter case this may be a nominated or domestic sub-contractor or a sub-contractor for a parcel of work under a Management Contract and in either case with or without design or manufacturing implications.

AUDITOR'S CHECKLIST FOR 2ND PARTY PROJECT ASSESSMENT

PROJECT AUDIT OBJECTIVES	Comments

1. Consider the particular purpose of this audit. Is it:-

   a) as part of the selection of a Design Team for a project?

   b) the appointment of a Contractor / Management Contractor for a project?

   c) the appointment of a Contractor or Sub-Contractor in a design and construct or design and manage role?

   d) the verification of a Contractor or a Sub-Contractor's performance as part of a domestic contract or parcel of work?

Data Required

2. The approach to the questionnaire and the audit of the company will be governed by the type of information required to fulfil these objectives and to establish the Quality Standards that the company will have to achieve in order to satisfy the specific project needs (conformity with requirements); and one of the criteria will be the proposed Form of Contract or Agreement.

3. Consider 2.PAQ and supporting documents. Obtain any additional support documents where possible before visit.

4. Establish the office(s) and personnel who are likely to be responsible for the specific project which is the subject of the Audit and ensure the location(s) visited are relevant. If manufacturing is part of the company's process it will be necessary to also visit the manufacturing plant and to 'Audit' their procedures. This will normally be undertaken subsequent to the visit to the relevant office(s).

5. Any work which has been noted in the questionnaire as a guide to Quality should be visited and any referees contacted prior to the visit to the office(s).

Inspect/Collect

6. Arrange to inspect (and sometimes, where particularly relevant, to collect) all printed documents which are available to assist in the audit.

eg:

letterheads	[ ]	invoices	[ ]
memo sheets	[ ]	telephone message pads	[ ]
timesheets	[ ]	job budget sheets	[ ]
brochures	[ ]	promotional documents	[ ]
order forms	[ ]	fax sheets	[ ]
petty cash vouchers	[ ]	V.O. forms	[ ]
payment advice slips	[ ]	job lists	[ ]
filing indexes	[ ]	checking in/out procedures	[ ]
staff lists	[ ]	works order forms	[ ]
telephone lists	[ ]	procedure notes	[ ]

site instruction sheets }
architect's   "   "   }   where appropriate
engineer's   "   "   }

etc, etc

Discussion

7. On visit, go through QMA with Senior Partner/Director and Managers who have completed the questionnaire.

8. Then proceed to question:-

How is post received - recorded? Despatched - recorded?

How is work allocated by Partners / Principals / Managers?

Are Partners / Principals / Managers involved in:-

design	[ ]
estimating	[ ]
valuation	[ ]
supervision, etc?	[ ]

Is there an Office Manager?   [ ]

What are his/her responsibilities?

Document Control	[ ]
Drawing register	[ ]
Job register, etc.	[ ]

9. Visit any location necessary to obtain an adequate cross-section of the nature of the company, the people who run it or are employed by it. This must include obtaining copies of all relevant forms and written and/or standard procedures / office manuals, etc; including printouts of any computer held data / drawings / specifications / schedules, etc, as well as information on organisation structure and responsibilities. Obtain floor plans of the premises and location of staff? (including any details of leases, agreements, etc) if thought likely to influence the preparation of Project Quality Plans, Objectives, Quality Policy, etc.

10. If little documentation or evidence of written procedures exists, establish those which are in existence, in whole or in part, by discussion with the relevant staff. Where written procedures do exist, check that they are understood and worked to by the staff.

Subjective Conclusions

11. Prepare (2nd Party) Audit Report for the Commissioning Client (guided by the purpose established at 1. above, and Contract Form at 2. above), recording the findings and observation of the findings, all with clear recommendations on further action to comply with Brief requirements. This may be a clear-cut recommendation to 'select' or 'not select' the particular company, or recommendation to select subject to the recommended modifications of procedures, etc. Where in agreement with the Client a report can also be submitted to the audited company outlining the findings and possibly the recommendations.

12. The Audit Report should be discussed with the commissioning client in detail and, where appropriate, may also be discussed with the audited company (who might also receive a precis of the report).

13. Finally, the audited company should be informed of the completion of the initial audit and of its outcome.

    If the company is selected then the basis, timing and methods of review audits during the course of the project should be indicated (See Module 214).

14. The Auditor might also make comments and suggestions of areas for any further consideration on the organisation/structure and on the APPROACH AND ATTITUDES of principals and staff. These should be made as points for discussion which will be held after the report has been issued to the commissioning authority, or given at a POST REPORT MEETING with the principals.

BIBLIOGRAPHY

Henri Fayol   1841—1925   With Constance Storrs. *General and*
(France)   *industrial management.* Pitman, London,
                                  1949.

Quotations from extract in *Organisation
theory*, edited by D.S. Pugh, Penguin,
London, 1990.

F.W. Taylor (USA) 1856—1917 *Scientific management.* Harper and Row,
New York, 1947.

Max Weber   1864—1920 *The theory of social and economic*
(Germany)   *organisation.* Free Press, New York,
1947.

Alfred P. Sloan   1875—1966 *My years with General Motors.*
(USA)   Doubleday, New York; Sidgwick and
Jackson, and Penguin, London, 1963,
1966, 1986.

Elton W. Mayo   1880—1949 *The social problems of an industrial*
(Austria)   *civilisation.* Routledge and Kegan Paul,
London, 1949.

Walter A. Shewhart 1891—1967 *Economic control of manufactured
production.* 1931.

W. Edward   1900—   *Out of the crisis.* Massachusetts Institute
Deming (USA)   of Technology Press and Cambridge
University Press, 1986, 1988.

With Mary Walton *The Deming
management method.* Dodd Mead, New
York, and Mercury, London, 1986,
1989.

With H.R. Neave *The Deming
dimension.* SPC, Knoxville, 1980.

Joseph M. Juran   1904—   *Quality control handbook.* 1951.
(Rumania)   *Juran on planning for quality.* Free Press,
New York, and Collier Macmillan,
London, 1988.

Douglas McGregor 1906—1964 *The human side of enterprise.* McGraw-
(USA)   Hill, New York, 1960, and Penguin,
London, 1987.

Leadership and motivation. Massachusetts
Institute of Technology Press, 1966.

Reg Revans (UK) 1907 — *Developing effective managers.* 1971.

Action learning. Blond and Briggs, London, 1979.

Abraham Maslow 1908 — 1970 *Motivation and personality.* Harper and Row, New York, 1970.

Peter Drucker (Austria) 1909 — *The practice of management.* Harper and Row, New York, 1954.

Managing for results. Heinemann, London, 1964, 1989.

Management tasks, responsibilities, practices. Heinemann, London and Harper and Row, New York, 1974.

The new realities. Heinemann and Mandarin, London, 1989, 1990.

Elliott Jaques (Canada) 1917 — *The changing culture of a factory.* Tavistock, London, 1951.

The measurement of responsibility. Tavistock, London, 1956.

Glacier project papers. Heinemann, London, 1965.

A general theory of bureaucracy. Heinemann, London, 1976.

Frederick Herzberg (USA) 1923 — *The motivation to work.* Wiley, New York, 1959.

Managerial choice: to be efficient and to be human. Dow Jones, Irwin, 1976.

John Humble (UK) 1925 — *Management by objectives.* McGraw-Hill, Maidenhead, 1971.

Ron Baden Hellard (UK) 1927 — *Management applied to architectural practices.* Goodwin, 1964.

Metric change — a management action plan. Kogan Page, London, 1971.

Training for change. Wellins, 1972.

Managing construction conflict. Longman, Harlow, 1988.

Charles Handy (UK) 1932 — *Understanding organisations.* Penguin, London, 1976.

The future of work. Basil Blackwell, Oxford, 1984.

The making of managers. Longman, London, 1988.

Inside organisation: 21 ideas for managers. BBC Books, London, 1990.

Edward de Bono (Malta) 1933 –

The use of lateral thinking. McGraw-Hill, Maidenhead, and Penguin, London, 1967.

The mechanism of mind. McGraw-Hill, Maidenhead, and Penguin, London, 1969.

Conflicts: a better way to resolve them. McGraw-Hill, Maidenhead, and Penguin, London, 1985.

I am right, you are wrong. Viking, London, 1990.

John Adair (UK) 1934 –

Effective leadership. Gower, Aldershot, 1983.

Effective teambuilding. Gower, Aldershot, 1986.

Not bosses but leaders. Talbot Adair, Guildford, 1988.

Understanding motivation. Talbot Adair, Guildford, 1990.

R.H. Waterman Jr. (USA) 1936 –

See **Tom Peters** for joint publications

Richard Schonberger (USA) 1937 –

World class manufacturing casebook. Free Press, New York, 1987.

Building a chain of customers. Free Press, New York, and Business Books, London, 1990.

Richard T. Pascale (UK) 1938 –

The art of Japanese management. Simon and Schuster, New York, and Allen Lane and Penguin, London, 1981, 1982, 1986.

Managing on the edge. Viking, london, 1990.

Henry Mintzberg 1939 –

Power in and around organisations. Prentice-Hall, New Jersey, 1983.

Mintzberg on management. Collier Macmillan, London, 1989.

Tom Peters (UK) 1942 –

With R.H. Waterman Jr. *In search of excellence.* Harper and Row, New York and London, 1982.

With R.H. Waterman Jr. *A passion for excellence.* Collins, London, 1987, 1988.

		With R.H. Waterman Jr. *Thriving on chaos*. Macmillan, London, 1987, 1988.
		With R.H. Waterman Jr. *The renewal factor*. Bantam, New York, 1987.
Rosabeth Moss Kanter (USA)	1943–	*Men and women of the corporation*. Basic, 1977.
		The change masters: corporate entrepreneurs at work. Allen and Unwin, London, 1984.
		When giants learn to dance. Simon and Schuster, New York and London, 1989.
Kenichi Ohmae (Japan)	1943–	*The mind of the strategist*. McGraw-Hill, New York, and Penguin, London, 1982, 1983.
		Triad power: the coming shape of global competition. Free Press, New York, 1985.
		The borderless world. Harper Business, New York, and Collins, London, 1990.
Michael Porter (USA)	1947–	*Competitive strategy: techniques for analysing industries and competitors*. Free Press, New York, 1980.
		Competitive advantage. Free Press, New York, 1980.
		Competition in global industries. Harvard Business School Press, Cambridge, Mass., 1986.

INDEX